*INTERNATIONAL SERIES OF MONOGRAPHS ON
PURE AND APPLIED BIOLOGY*

Division: **ZOOLOGY**
General Editor: G. A. Kerkut

Volume 15

THE PHYSIOLOGY OF
EARTHWORMS

## OTHER TITLES IN THE ZOOLOGY DIVISION

*General Editor:* G. A. KERKUT

- Vol. 1. RAVEN – *An Outline of Developmental Physiology*
- Vol. 2. RAVEN – *Morphogenesis: The Analysis of Molluscan Development*
- Vol. 3. SAVORY – *Instinctive Living*
- Vol. 4. KERKUT – *Implications of Evolution*
- Vol. 5. TARTAR – *The Biology of Stentor*
- Vol. 6. JENKINS – *Animal Hormones – A Comparative Survey*
- Vol. 7. CORLISS – *The Ciliated Protozoa*
- Vol. 8. GEORGE – *The Brain as a Computer*
- Vol. 9. ARTHUR – *Ticks and Disease*
- Vol. 10. RAVEN – *Oogenesis*
- Vol. 11. MANN – *Leeches (Hirudinea)*
- Vol. 12. SLEIGH – *Biology of Cilia and Flagella*
- Vol. 13. PITELKA – *Electron-Microscopic Structure of Protozoa*
- Vol. 14. FINGERMAN – *The Control of Chromatophores*

## OTHER DIVISIONS IN THE SERIES ON PURE AND APPLIED BIOLOGY

BIOCHEMISTRY

BOTANY

MODERN TRENDS IN PHYSIOLOGICAL SCIENCES

PLANT PHYSIOLOGY

# THE PHYSIOLOGY OF EARTHWORMS

BY

M. S. LAVERACK

*Gatty Marine Laboratory, The University,
St. Andrews, Fife*

A Pergamon Press Book

THE MACMILLAN COMPANY
NEW YORK
1963

THE MACMILLAN COMPANY
60 Fifth Avenue
New York 11, N.Y.

This book is distributed by
THE MACMILLAN COMPANY – NEW YORK
pursuant to a special arrangement with
PERGAMON PRESS LIMITED
Oxford, England

Copyright © 1963
PERGAMON PRESS LTD.

Library of Congress Card Number: 62-22103

*Set in 11 on 12 pt. Imprint and printed in Great Britain
at the Alden Press, Oxford*

# CONTENTS

|  |  | PAGE |
|---|---|---|
|  | Preface | vii |
|  | Acknowledgments | ix |
| I | Biochemical Architecture | 1 |
| II | Digestion and Metabolism | 18 |
| III | Calciferous Glands | 24 |
| IV | The Axial Field | 36 |
| V | Nitrogenous Excretion | 45 |
| VI | Water Relations | 69 |
| VII | Respiration | 83 |
| VIII | The Physiology of Regeneration | 118 |
| IX | Neurosecretion | 128 |
| X | Nervous System | 138 |
| XI | Behaviour | 171 |
|  | References | 185 |
|  | Index | 202 |

# PREFACE

It will be found that the work reported in this book leans heavily upon earthworm species whilst other oligochaetes such as the fresh-water species are but little mentioned. Little is known of many aspects of the physiology of fresh-water oligochaetes and it is hoped that interesting problems in the field will be indicated in the present work.

Throughout the book three species occur again and again, *Lumbricus terrestris*, *Eisenia foetida* and *Allolobophora longa* and these have generally been abbreviated to *L. terrestris*, *E. foetida* and *A. longa* to avoid tedious repetition. All other species are given their full title unless described more than once in close proximity.

I have endeavoured to cover the major advances in oligochaete physiology since the publication of Stephenson's major work "The Oligochaeta" in 1930. This is an excellent source book for most early work. Advances in modern techniques have rendered much of this early work out of date, however, and I hope that this will be remedied by the present review.

# ACKNOWLEDGMENTS

I AM indebted to Messrs. H. R. Bustard, R. A. Chapman and J. Scholes, and to Dr. J. E. Satchell for their many helpful comments and criticisms. Any faults and flaws remaining are in no way due to them and I thank them for their assistance.

Thanks are also due to the Editors and Publishers of the following journals and publications for their permission to utilize figures and tables reproduced in the text, and to Dr. J. N. Parle for allowing me to reproduce the substance of a letter to Dr. J. E. Satchell. Académie Royale de Belgique; *Annales des Sciences Naturelles Zoologie*; *Autralian Journal of Experimental Biology and Medical Science*; *Australian Journal of Science*; *Biochemical Journal*; *Biochimica et Biophysica Acta*; *Biological Reviews*; *Compte Rendus de la Societé de Biologie*; Gustave Fischer Verlag; *Journal of Biological Chemistry*; *Journal of Biophysical and Biochemical Cytology*; *Journal of Cellular and Comparative Physiology*; *Journal of Experimental Biology*; *Journal of Experimental Zoology*; *Journal of Neurophysiology*; Masson et Cie; *Nature*; Oxford University Press; *Physiological Zoology*; Rockefeller Institute; Unione Zoologica Italiana; University of Chicago Press; Wildlife Society; Wistar Institute; Zoologische Jahrbücher. Also my thanks are due to Drs. J. Hanson, I. Linn and K. M. Rudall for bringing material to my notice.

Last but not least, my thanks go to Miss V. R. Taylor and to my wife for their aid in the preparation of the manuscript for publication.

CHAPTER I

# BIOCHEMICAL ARCHITECTURE

IN THE world of biology the chemical elements are arranged in complex patterns to form the body of plants or animals. The combination of carbon, nitrogen, phosphorus, sulphur, oxygen, hydrogen and other elements to form carbohydrates, amino-acids, peptides, proteins, and other organic compounds provides the basis for the study of biochemistry.

Any animal body is composed of such substances, together with other materials such as water and mineral salts. The analysis of biological substances has recently been given a great impetus by the application of modern techniques such as chromatography, electrophoresis, flame photometry, and electron microscopy. These methods have both simplified and expanded the task of those interested in the structure of the tissues of animals. The recent discoveries of Sanger on the amino-acid sequence in insulin, and of Perutz and Kendrew on the molecular constitution of haemoglobin and myoglobin indicate the ultimate refinements of such analytical methods.

A great deal of the pioneer work in analysing bodily composition has been confined to the mammals because of the ease of obtaining raw materials in large enough quantities for detection of substances in small quantity, even by the most delicate instrumentation. It is only comparatively recently that the new techniques have been applied to the study of invertebrate structures. It is to be hoped that this embryonic interest will continue to thrive for the results are indicative of very interesting facts waiting to be discovered. This is exemplified in the oligochaetes by the analysis of the compound acting as a phosphagen in the earthworm, lombricine, in contradiction to the long held view that arginine phosphate is the phosphagen common to invertebrate animals.

Most of the work appertaining to the biochemical structure of earthworms, however, was performed before the methods mentioned above were in use. It is none the less interesting, as it is hoped to show in the pages that follow.

Undoubtedly the major chemical component of the oligochaete body, as that of any animal body, is water. The amount of water present in the earthworm has been variously estimated at 81·3% (Durchon and Lafon, 1951), 84·1% (Schmidt, 1927), 84·8% (Roots, 1956) and 88·0% (Jackson, 1926) of the entire body weight. This water is distributed in the coelom, blood vascular system and the tissues themselves. In both terrestrial and fresh-water oligochaetes the proportion of body water is always subject to change, often rapid, due in the main to desiccation in the land forms, flooding under osmotic stress in the fresh-water types, and the secretion of mucus by both classes.

Earthworms of the species *Lumbricus terrestris* and *Allolobophora caliginosa* show regional differences in the water content of the integument, the anterior portion containing more than the posterior extremity. The capacity of the body wall to take up water also varies from place to place, and a minimum supply of calcium ions, an important factor in membrane permeability, is necessary to prevent flooding of the interior with water which passes rapidly through the body wall from the external environment. Hydrophobic substances such as lipids also play a part in maintaining a water barrier in the integument, for if they are removed by alcohol treatment water uptake, particularly in the anterior regions, is very rapid (Kopenhaver, 1937).

Variations in integumentary water contents have also been described for two Indian species, *Pheretima posthuma* and *Lampito* (*Megascolex*) *mauritii*, in both of which the amount of water, estimated in pre-clitellar, post-clitellar and rectal fragments, increases from the anterior to posterior regions. At the same time it is found that the absolute quantity of water present in these Indian earthworms is lower than that reported for temperate species. This may indicate that the drier soils of the tropics influence the water content of the integument, although no information is given as to what happens to the water content when rainy monsoon conditions prevail (Tandan, 1951). It has been shown that in temperate countries earthworms obtained fresh from the soil are rarely fully hydrated and indeed may absorb up to 15%

of their body weight during 5 hours immersion in water (Wolf, 1940). In the often very dry soils of the tropics it is possible that the normal dehydration of earthworms proceeds even further.

It is evident from the foregoing that water is able to cross the boundary formed by the skin with comparative ease. The ability to survive although the water content of the body shows considerable and rapid fluctuations is evidently of great value, but little work has yet been directed towards an elucidation of the mechanisms by which bodily functions are maintained under conditions of water stress, either ingoing or outgoing.

Earthworms show great tolerance towards water loss, and can recover from drastic dehydration. If water is lost rapidly a deficit of 43–50% body weight in 5–9 hours can be withstood; Jackson (1926), and Schmidt (1927) kept earthworms alive after they had lost 30% body weight in 5¾ hours. A more protracted drying period, as used by Roots (1956), of 20–24 hours can lead to a greater loss of up to 57–59·7% of body weight and the animals, *Lumbricus terrestris*, still remain alive. Under the same conditions a smaller species, *A. chlorotica*, loses 50% of its body weight in 3 hours. A loss of 60% of body weight, corresponding to 70% of the total water content of *L. terrestris*, and 75% of total water of *A. chlorotica*, can be tolerated. The ability of earthworms to withstand such great losses of water is of aid in maintaining field populations under exceptionally arid conditions (Zicsi, 1958).

## Chemical Composition of the Body

*Whole Body*

The average dry weight of earthworms is about 15–20% of the fresh weight. This represents the entire carcass weight after complete dehydration. The chemical constitution, protein, amino-acids, carbohydrates, etc., of the carcass has been investigated for only a few oligochaete species.

Protein accounts for the largest fraction of the dry weight. It has been estimated variously (see Table 1) at between 53·5% and 71·5% of the total dry weight of *Lumbricus terrestris*. This indicates a wide variation in the body proteins, and this variation is also shown by analysis of other types of biochemical compounds. For example Durchon and Lafon (1951) found a lipid content amounting to 17·3% of total dry weight, and an ash residue of 9·2%. These figures contrast with those of French, Liscinsky and Miller (1957) who obtained figures of 6·07% and 23·07% respectively. Lawrence and Millar (1945) give an even lower figure still for lipid contents, 1·5% of dry weight. Table 1 summarizes the analysis as known at present.

Two smaller species, *Lumbricus rubellus* and *Eisenia rosea* have a similar proportion of dry weight; 16·38% which consists of 61·3% protein, 17% carbohydrates, 4·5% fat and only 15% remains as ash residue, and mineral salts (French, Liscinsky and Miller, 1957). No figures are as yet available for a fresh-water oligochaete such as *Tubifex tubifex*.

TABLE 1

CHEMICAL CONTENT OF *L. Terrestris*

|  | Durchon and Lafon 1951 | French, Liscinsky and Miller 1957 | Lawrence and Millar 1945 |
|---|---|---|---|
| Dry wt. | 18·7% | 17·38% |  |
| Lipids | 17·3% | 6·07% | 1·5% |
| Protein | 56·9% | 53·5% | 62–71·5% |
| Ash | 9·2% | 23·07% |  |
| Glycogen | 11·0% | 17·42% (carbohydrate) |  |

*Fats*

Individual classes of biochemical substances have received a little attention. For example a review of the fats and sterols represented among invertebrates (Bergmann, 1949) included details of four oligochaetes. Oil extracted from *Perichaeta communissima* accounts for 2·3% of the total dry weight and contains 31·2% unsaponifiable material. Similar oil extracted from *L. terrestris* contains 36·7% unsaponifiable material, *L. rubellus* 21·3% and *Tubifex tubifex* 10·5%. De Waele demonstrated that cholesterol is the major sterol in *L. terrestris* and Bock and Wette have found that more than 20% of the original extractable oil consists of a provitamin D (ergosterol) and other poorly defined sterols (Bergmann, 1949).

*Amines*

Among nitrogen-containing substances much research has been carried out on the amino-acids represented in many animals and yet apart from one very significant study little has been published using the highly refined methods of recent years on oligochaete species. Early chemical analysis of homogenates from *L. terrestris* shows that adenine, lysine, leucine, tyrosine, betaine, choline and lactic acid are present (Ackermann and Kutscher, 1922). Likewise

TABLE 2

AMINO-ACID CONTENT OF EARTHWORM HAEMOGLOBIN

|  | % |
|---|---|
| Tryptophan | 4·41 |
| Tyrosine | 3·47 |
| Arginine | 10·07 |
| Histidine | 4·68 |
| Cystine | 1·41 |
| Lysine | 1·73 |

the haemoglobin of the blood of *L. terrestris* has been analysed and contains at least six amino-acid residues (Table 2, Florkin, 1955).

Paper chromatographic techniques would confirm and expand these observations when applied to a study of the blood, coelomic fluid and muscle homogenates. The interest of such investigation is particularly pointed when one considers that the first conclusive report of a D-enantiomorph amine in animal tissue has recently been made with regard to earthworms. Thoai and Robin (1954) hydrolysed muscle homogenates in a study of the guanidine derivatives and phosphagens of annelids. Lombricine (see later), the phosphagen of earthworms, contains the serine moiety and this was shown to occur as the D-form by Beatty, Magrath and Ennor (1959). Later work has elucidated aspects of the metabolism of this compound.

The fullest amino-acid analysis yet available refers to two

primitive oligochaetes *Aeolosoma hemprichi* and *A. variegatum*. Eleven amino-acids have been identified in hydrolysates of both these species: ten are common to both species, alanine, aspartic acid, glutamic acid, glycine, leucine (or iso-leucine), lysine, proline, serine, tyrosine and valine. In addition threonine is found only in *A. variegatum* and methionine only in *A. hemprichi* (Auclair, Herlant-Meewis and Demers, 1951).

*Muscles and Myosin*

The hydrolysis of the whole body, as in *A. hemprichi*, or of the body wall musculature as performed by Thoai and Robin (1954) breaks down the component structure of the tissues. In particular it destroys the continuity of the muscle tissues converting protein into the constituent amino-acids.

The muscle layers of the body wall of *L. terrestris* are composed of smooth fibres that lack the striations of typical vertebrate skeletal muscle, but nevertheless have cross stripes. The individual fibres have the shape of a ribbon with tapered ends. The width of the ribbon is about 20 $\mu$ and the thickness between 2 and 5 $\mu$. The contractile part of the fibre is covered with a thin layer of undifferentiated sarcoplasm to which the mitochondria are confined, and which contains a single nucleus. The fibrils, lying with one edge at the surface of the fibre, and the other towards the interior, are also ribbon-shaped. When the fibre is viewed in plan the two sets of fibrils, one on either side of the ribbon, are found to show an angle between them, never being parallel, and the value of this changes depending upon whether or not the fibre is contracted (Fig. 1a).

The fibrils show the same structure along their length but no A or I bands are differentiated. There are conspicuous filaments lying parallel with the long axis of the fibre, and there are approximately 100 filaments in a typical cross-section through a fibre. They appear solid and circular in cross-section with diameters ranging from 120–300 Å.

Between the fibres other structures are found. Each fibril bears stripes upon one surface and bridges external from the stripes to the adjacent fibril, but the bridges do not connect stripe to stripe, rather they connect stripe to the stripeless surface of the next fibril (Fig. 1b, Hanson, 1957). Hanson and Lowy (1960) indicate that the organization of earthworm muscle is that of a sarcoplasmic reticulum. The arrangement is of smooth muscle fibres with helically arranged myofibrils.

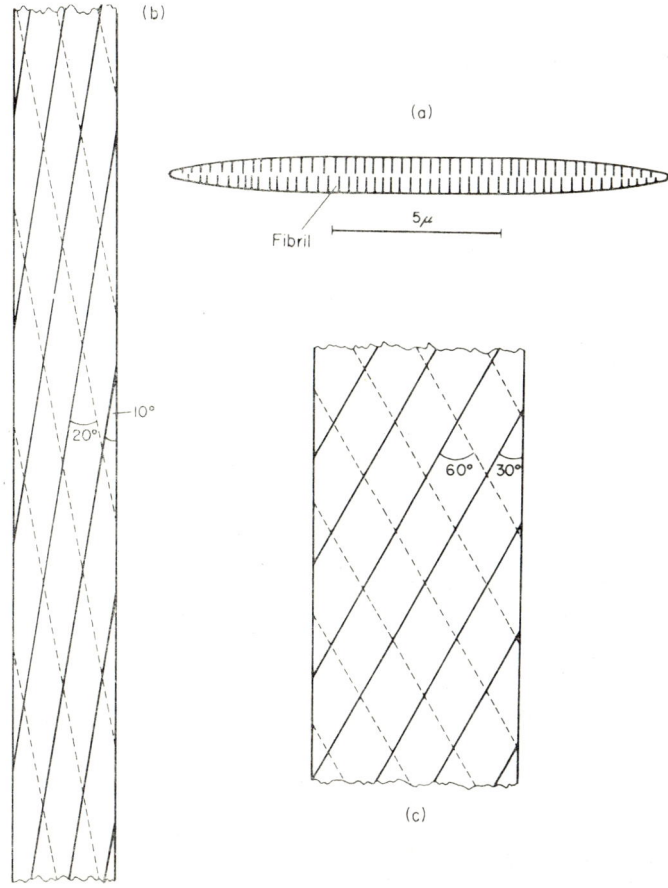

Fig. 1a. Diagram of a transverse section through a longitudinal muscle fibre in the body wall of an earthworm (a). The fibre is ribbon-shaped (b). The angles between the fibres are indicated in the stretched and contracted conditions (c).

The proteins of this organized tissue have been separated by electrophoretic means. Five fractions in all have been noted, one being haemoglobin presumably from blood contained within the muscle layers. Of the other fractions two are albumen and two are myosin, termed $\alpha$ and $\beta$ and which are claimed to be equivalent to those of vertebrates (Godeaux, 1954) (Fig. 2).

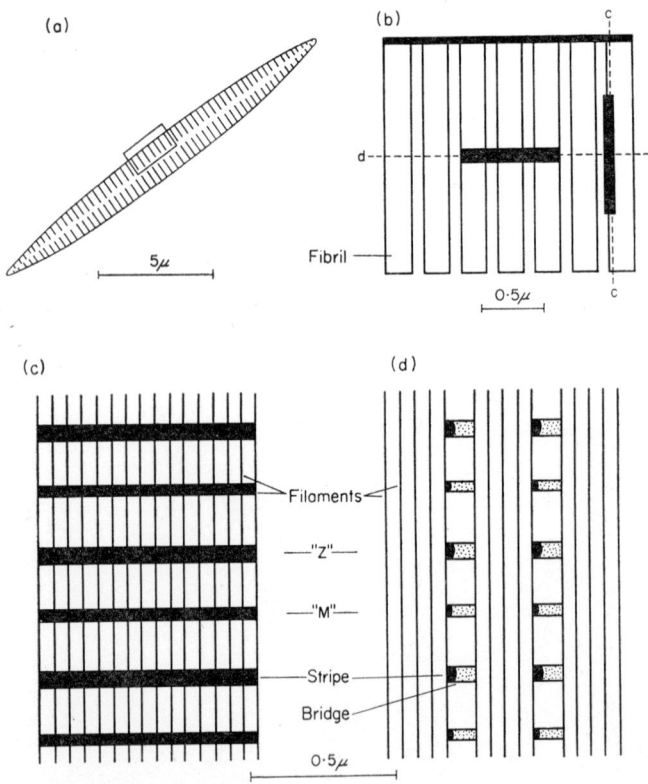

Fig. 1b. Diagrams to show the arrangement of stripes and bridges in a muscle fibre. (a) A fibre cut in transverse section. (b) An enlarged view of the area marked in (a). Black regions marked c–c, d–d are those further enlarged in Figs. (c) and (d) (from Hanson, 1957).

The similarity of myosin obtained by fractionation from an earthworm, *Pheretima communissima*, with that of striated insect and vertebrate muscle has been pointed out by Maruyama and Kominz (1959). The major differences lie in the viscosity and flow birefringence of the preparations. Ultracentrifugation separates two types of myosin A and B ($\equiv \alpha, \beta$, Godeaux, 1954) and these are somewhat contaminated by tropomyosin A. The amount of myosin A ($\alpha$) is increased and that of myosin B ($\beta$) decreased, by the addition of ATP. Myosin B shows typical ATP-ase properties and calcium ions act as a co-factor. Tropomyosin A has been

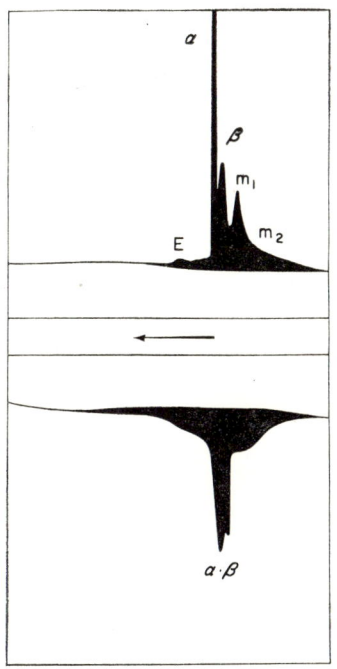

Fig. 2. Electrophoretic pattern, pH 7·45, of the body wall of *L. terrestris*; E = erythrocruorin; $\alpha$ and $\beta$ = myosins; $m_1$ and $m_2$ = albumens (from Godeaux, 1954).

demonstrated by Kominz, Saad and Laki (1957), and filaments resembling paramyosin by Hanson and Lowy (1960).

Flow birefringence measurements reveal that the molecular length of myosin B is 1·3–0·6 $\mu$ in the absence of ATP, and 0·64–0·43 $\mu$ in the presence of ATP and this is shorter than the length obtained from striated muscles of insects. Myosin A, on the other hand, has a molecular length of 1700 Å which is very similar to that of insect striated muscle. It is suggested that the degree of polymerization of myosin B is less than that of striated muscle, thus accounting for the shorter molecular length. It is not yet known whether actin is involved in the polymer formation as is the case in vertebrates (Marnyama and Kominz, 1959).

*Pigmentation*

The outer layer or integument of some species of earthworm is often heavily pigmented, a purple-brown substance being deposited in the tissues throughout life (Stephenson, 1930). This pigment is easily extractable from the integument and in solution exhibits a fine red fluorescence under ultra-violet illumination.

In 1886 MacMunn identified this red fluorescent compound as a porphyrin and suggested that, in common with pigments obtained from *Asterias rubens*, *Limax flavus* and *Arion ater* this substance was haematoporphyrin, a breakdown product of haemoglobin. This opinion was supported by Kobayashi (1928) who isolated 20 mg of porphyrin hydrochloride from 21 kg of *E. foetida*. The absorption spectrum of this pure material had peaks at 518, 530, 559 and 607 m$\mu$ (in acid alcohol) and 502, 547, 583 and 622 m$\mu$ (in alkaline alcohol). Kobayashi considered this to be very similar to the spectrum of haematoporphyrin.

Dhéré (1932) on the other hand, found that the pigment is soluble in pyridine to give a red fluorescent solution having a spectral emission under u.v. that was identical to protoporphyrin. Fischer isolated a specimen of this material and identified it as Kammerer's Porphyrin, a substance now known to be the same as protoporphyrin (see Vannotti, 1954).

Haematoporphyrin is now known to be absent in nature, being found only in laboratory degradations of haemoglobin. The opinions of MacMunn (1886) with regard to the identity of pigments in *Arion*, *Asterias* and *Limax* have been shown to be erroneous by modern techniques (Kennedy, 1959; Kennedy and Vevers, 1953), and his ideas on *Lumbricus* were also wrong, as indicated by Dhéré and Fischer and recently confirmed by chromatography by Laverack (1960a).

Laverack (1960a) found that a red fluorescent pigment of the body wall was extracted by ether containing acetic acid, giving a brown solution. The solution fluoresces red, but the body wall still retained a considerable amount of pigment which was not extracted by prolonged ether : acetic acid treatment. It was easily released by methanol : sulphuric acid (Kennedy and Vevers, 1953) to give a red-violet solution which fluoresces intensely. The two fractions obtained show distinctive behaviour on paper and column chromatographs but possess the same absorption spectra, having absorption peaks at 503, 541, 576 and 632 m$\mu$ (in chloroform solution) and these peaks correspond very closely with those given by authentic protoporphyrin. From this evidence the two pigments have been identified as protoporphyrin and protoporphyrin methyl ester respectively (Fig. 3). Small amounts of a third red

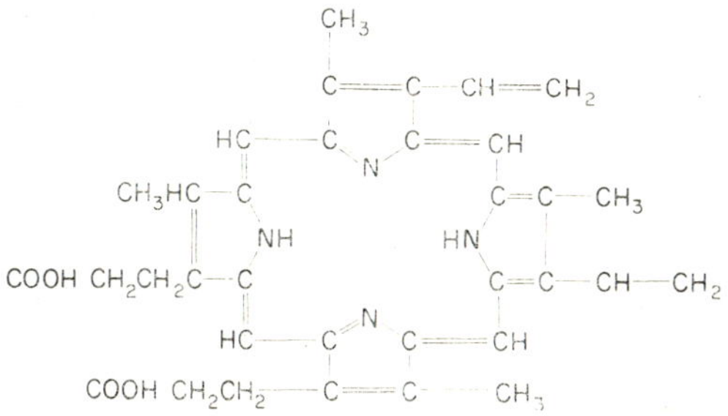

Fig. 3. Formula of protoporphyrin.

fluorescent substance have also been seen, probably representing an intermediate product.

These red fluorescent porphyrin compounds are associated with a second pigment, deep brown in colour, which remains unidentified.

Kalmus, Satchell and Bowen (1955) have described investigations into the composition of a pigment obtained from a green version of *Allolobophora chlorotica*.

*Mucus*

A property of many of the cells of the epidermis of earthworms is the secretion of mucus (Fig. 4). This material is copiously produced in response to various noxious stimuli and may act as a buffer defence system against such stimuli. Although no detailed study has yet been made of the chemical composition of the mucus it is probably a mucoprotein since Needham (1957) states that about 50% of the nitrogen lost from the body is in the form of mucus protein. Ewer and Hanson (1945) have shown that various epidermal cells stain with mucicarmine and are metachromatic with thionin. In the light of recent research on mucoproteins, on the occurrence of chondroitin sulphate, and on the phenomenon of metachromasia further knowledge of this secretion should not be long delayed.

*Cuticle*

The epidermal cells also secrete the cuticle which extends as a

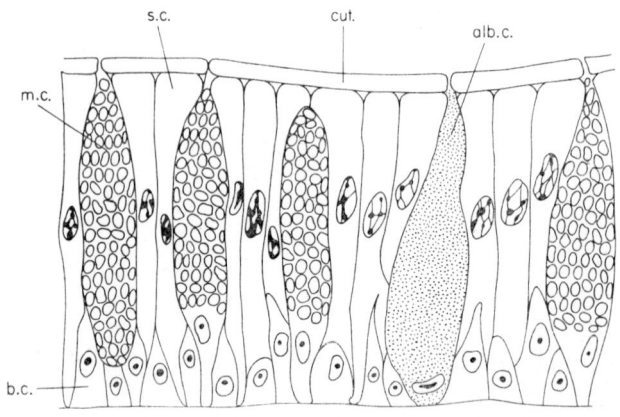

FIG. 4. Section of epidermis of *Lumbricus terrestris*. alb.c. albuminous gland cell; b.c. basal cell; cut. cuticle; m.c. mucous cells; s.c. supporting cells (from Stephenson, 1930).

thin layer over the whole of the external body surface, and also the fore and hind gut regions. It is an inert substance and can be easily separated from the body wall, particularly if the animals are killed by immersion in ether (Watson, 1958).

In its chemical and physical properties the earthworm cuticle is similar, but not identical, to the collagen of vertebrates. The first demonstration that the cuticle is a typical fibrous protein was made by Picken, Pryor and Swann (1947). Using X-ray diffraction techniques they showed that the fibrous protein is embedded in a matrix such that the fibres are orientated with one another. They cross, making an angle of 90° with one another, and lie at 45° with reference to the long axis which itself lies in the plane of the cuticle. Millard and Rudall (1960) have shown that the deepest regions of the cuticle, nearest the secreting epithelial cells, are probably areas in which the filaments are still growing. In the outermost layers there is a sudden decrease in size of filaments as if their substance was being resorbed. "Papillae" arising from the surface of the epithelial cells contain tube-like prolongations of the cell interior passing out to the plasma membrane. These may be the points of secretion of the cuticular filament substance.

The collagenous nature of the earthworm cuticle has been confirmed many times since (Rudall, 1955; Watson and Smith, 1956; Watson, 1958), but there is at least one important difference compared with vertebrate collagen: there is no periodicity of the fibrils at 640 Å intervals in either *L. terrestris* (Watson and Silvester, 1959), or *A. longa* (Reed and Rudall, 1948) (Fig. 5).

The cuticle is highly proteinaceous in character, some 80% of the total being composed of protein whilst the remaining 20% is made up of polysaccharides, galactose, pentoses and hexosamines (Watson, 1958). Nitrogen accounts for 14·6% of the cuticle weight and at least 16 amino-acids are represented in the protein. The amino-acids of the cuticle of two species, *L. terrestris* and *A. longa* are listed in Table 3 where they are compared with a typical vertebrate collagen, that of ox-hide (Watson and Smith, 1956; Watson, 1958; and Singleton, 1957). Amines with non-polar side chains e.g. glycine, are present in quantities similar to those in vertebrate collagen, but arginine, histidine, and lysine with basic side chains are less abundant than in vertebrate material, whilst aspartic acid, glutamic acid and serine (D-serine?) with acidic side chains are present in greater quantities than in the vertebrate counterpart. There is no trace of hydroxylysine in earthworm cuticle.

The gizzard of *L. terrestris* is lined with a cuticle. This can be removed and Rudall (1955) subjected this material to X-ray analysis, obtaining patterns similar to those of chaetae (see p. 15). Cuticle from this situation is not stable in hot dilute alkali, does not crystallize, and though chitin-like is not in the α-chitin form. The presence of a chitin would be shown by a fibrillar structure, but van Gansen (1959b, 1960) using similar material finds no evidence for the belief that the gizzard lining is fibrillar. Chemical tests with orcein, resorcin, van Gieson stain, benzidine and Gomori stain

TABLE 3

AMINO-ACID COMPOSITION OF EARTHWORM CUTICLE (TWO SPECIES) COMPARED WITH THAT OF OXHIDE

| Species | L. terrestris | | L. terrestris | | A. longa | | Oxhide |
|---|---|---|---|---|---|---|---|
| Authors | Watson and Smith (1956) | | Watson (1958) | | Singleton (1957) | | Watson (1958) |
| Amines | g/100g | N as % total N | g/100g | N % | g/100g | N % | N as % total N |
| Total N | 14·6 | | 14·6 | 25·0 | 15·3 | 30·0 | 26·6 |
| Glycine | | 24·6 | 19·6 | 8·31 | | 6·8 | 8·72 |
| Alanine | | | 7·72 | 2·73 | | 2·8 | 2·14 |
| Leucine | | | 3·74 | 1·42 | | 2·8 | 1·08 |
| Isoleucine | | | 1·94 | 1·77 | | 2·9 | 1·58 |
| Valine | | | 2·15 | 1·11 | | 2·9 | 1·07 |
| Phenylalanine | | | 1·92 | 0·33 | | 0·8 | 0·41 |
| Tyrosine | | | 0·63 | 8·5 | | 8·4 | 3·06 |
| Serine | | 8·1 | 9·3 | 5·0 | | 4·4 | 1·43 |
| Threonine | | 5·0 | 6·2 | 0·0 | | | |
| Cystine | | | 0·0 | 0·0 | | | 0·49 |
| Methionine | | | 0·0 | 1·17 | | 3·1 | 9·43 |
| Proline | | 1·2 | 1·41 | 12·8 | | 14·6 | 7·37 |
| Hydroxyproline | | 12·6 | 17·5 | 7·04 | | | 14·22 |
| Arginine | | 6·5 | 3·21 | 0·15 | | | 1·02 |
| Histidine | | | 0·08 | 0·0 | | | 1·7 |
| Hydroxylysine | | | 0·0 | 2·58 | | 2·8 | 4·08 |
| Lysine | | 2·5 | 1·95 | 4·95 | | 4·4 | 3·93 |
| Arspatic acid | | | 6·88 | 6·63 | | 7·8 | 5·69 |
| Glutamic | | | 10·17 | 0·94 | 0·37 | | |
| Hexosamine | | | 1·76 | 4·9 | | | |
| Hexose | | | 14·0 | 4·9 | | | |
| Pentose | | | 2·4 | | | | |

(a)

FIG. 5. The structure of the earthworm cuticle. The cuticle is removed from the animal and subjected to layer-stripping to expose the structure at various depths in the cuticle. (a) Middle layers of earthworm cuticle. (b) Underneath surface of the external corpuscle layer of the cuticle (by kind permission of Dr. K. M. Rudall).

Fig. 5 (b.)

all point rather towards an affinity with vertebrate elastin, in distinct contrast to the collagen-like ectodermal cuticle. The earthworm elastin differs from elastin originating in vertebrates in that it is not birefringent or auto-fluorescent. Hydrolysis leads to the appearance of only four identified amino-acids, proline, hydroxyproline, valine and glycine. Van Gansen (1960) suggests that the lining of the gizzard is an elastin secreted from the cells of the organ in a manner comparable to the secreted collagen of the epidermal cuticle.

*Chaetae*

Closely associated with the cuticle and an integral part of the surface presented by the earthworm to the outside world, are the chaetae. There are usually four pairs of these small hard structures in each segment and they are secreted by glands in the dermal layers of the skin. They are used in locomotion as the animal passes along its burrow, and serve as an efficient anchor to maintain contact with the burrow when the anterior end is extended on the surface of the ground.

Chaetae can be isolated from the body by digestion with hot dilute alkali and when oriented parallel to one another show X-ray diffraction patterns very like those of $\beta$-chitin from vertebrates. The two substances are not identical but the similarities are greater than are the similarities between earthworm cuticle and vertebrate $\alpha$-chitin. Annelid chitin differs in small ways from typical arthropod chitin (Rudall, 1955). The chitin is associated with protein in the cuticle.

The chaetae of *A. longa* are amber in colour and Dennell (1949) suggests that this colour is due to tanning by an orthoquinone, as is the case with the browning of insect cuticle. Dennell (1949) also states that the chaetae dissolve at the proximal end in boiling concentrated KOH, followed by cold concentrated HCl, to leave a thin sheath or covering cap. This sheath covers precisely that part of the chaetae which is extended in locomotion and its resistance to HCl suggests a resemblance to the thin epicuticle of an insect.

*Phosphagen*

The energy cycles of animals are continually metabolizing and synthesizing high energy phosphate bonds which are mobilized through the agency of ATP. Excess phosphate bonds are stored in a form bound to a substance known as phosphagen. In the vertebrates this is exclusively creatine phosphate. Until quite recently (see review by Ennor and Morrison, 1958) it was thought that the vast majority of invertebrates relied on arginine phosphate as their phosphagen. This phenomenon was thought to throw light on evolutionary processes because such forms as echinoderms and enteropneusts apparently possess both creatine and arginine phosphate in their tissues (for table see Kerkut 1960, p. 112).

Within the last few years, however, it has become increasingly obvious that the invertebrates are not a homogeneous group and that a number of phosphagens other than arginine phosphate are to be found. Three different phosphagen substances have now been described in annelids: taurocyamine in *Arenicola*, glycocyamine in *Nereis* and lombricine in *Lumbricus* (Thoai and Robin, 1954).

$$OH.PO\begin{cases} OCH_2.CH_2.NH.C(NH).NH_2 \\ O.CH_2.CH(NH_2).COOH \end{cases}$$

FIG. 6. Formula of lombricine.

Arginine phosphate may be present but is not active as a phosphagen.

Earthworm muscles hydrolysed in 6N HCl for 8 hours at 110 °C break down into their constituent parts, and among these Thoai and Robin (1954) found the amino-acid serine and a number of related guanidino derivatives. These included guanidoethyl-seryl-phospho-diester, which they named lombricine.

This substance is found alone in the muscles of the earthworm, but in conjunction with arginine in the gut wall. The associated phosphagen phosphoguanidyl-ethyl-seryl-phosphate, is also present in the tissues (Thoai and Robin, 1954, Fig. 6). Phosphagen is at a concentration of 6·1 $\mu$moles/g. ATP amounts to 2·8 $\mu$moles/g and ADP 0·7 $\mu$moles/g (Rey, 1956).

Further details regarding the isolation, structure, biological synthesis and precursors of lombricine are to be found on p. 114.

*Summary*

The oligochaete body contains much water (about 80–85%)

which is present as blood, coelomic fluid and tissue water. Considerable interchange of water occurs across the body wall which shows regional variations in permeability. Surprising tolerance to uptake and loss of water is exhibited.

Details of the chemical composition of the body await further investigation by the most modern analytical techniques. Figures are available for the amounts of protein, fat, carbohydrate and mineral ash, but only one or two studies have given details of the precise biochemical nature of the compounds present. Two recent studies of the subject have shown that the collagen-like cuticle of the earthworm bears resemblances to, but differs from, the collagen of vertebrates and that pigmentation is partly due to a mixture of porphyrin compounds.

Most significant of biochemical results on the structure of earthworms has been the elucidation of the phosphagen compound. This is not arginine phosphate but lombricine. This substance contains the D-serine moiety: the first fully substantiated report of a D-amino acid to be found in the animal kingdom.

CHAPTER II

# DIGESTION AND METABOLISM

ANIMALS obtain the energy necessary for the maintenance of their bodily processes, such as muscle contraction, gland secretion, nervous activity and so on by the breakdown of complex organic substances obtained in their diet. This diet may be taken in a variety of ways: grazing, filter feeding, predation or burrowing, but in all cases the food must eventually be broken down by the animal ingesting it, i.e. digested, then absorbed and finally utilized i.e. metabolized. The mechanism of synthesis of animal protein takes place using only the simple building blocks of amino-acids, carbohydrates, etc., that can be absorbed by the gut. The exceptions to these generalizations are, of course, parasitic forms in which the alimentary system may have disappeared altogether.

The majority of oligochaetes are omnivores. That is to say they are non-specific eaters. Non-selective would be too widespread a concept since it has been shown that particular types of leaf are eaten in preference to others, at least by the earthworm (Mangold, 1951; Wittich, 1953), indicating that the animal evidently recognizes differences in its diet, presumably by sensory means. Earthworms (*L. terrestris*) come to the surface of the ground at night, searching mainly in the area surrounding the burrow entrance for food and drawing suitable leaves into the mouth of the burrow. When the litter layer is freshly fallen, leaves are grazed in a different preference order to the regime existing when decomposition has been proceeding for some while (see Chapter XI). This is possibly explicable in terms of chemical changes occurring under such conditions.

It is obvious that during the passage of earthworms through soil, and the making of burrows, much soil passes along the gut. This soil contains not only plant matter, but also the decomposing remains of animals large and small, living protozoa, rotifers and sundry other minute animal forms, and these will all be ingested adding their bulk to the food foraged on the surface of the ground. A number of fresh-water oligochaetes e.g. *Tubifex* are also omnivorous in the muds of lake and river bottoms, though probably a considerable proportion of their intake will be bacteria, since they live in situations in which oxygen is often at a premium. A few oligochaetes are carnivorous, consuming rotifers, fish, frogs, flies and other worms (Stephenson, 1930), and the Branchiobdellidae are external parasites on crayfish.

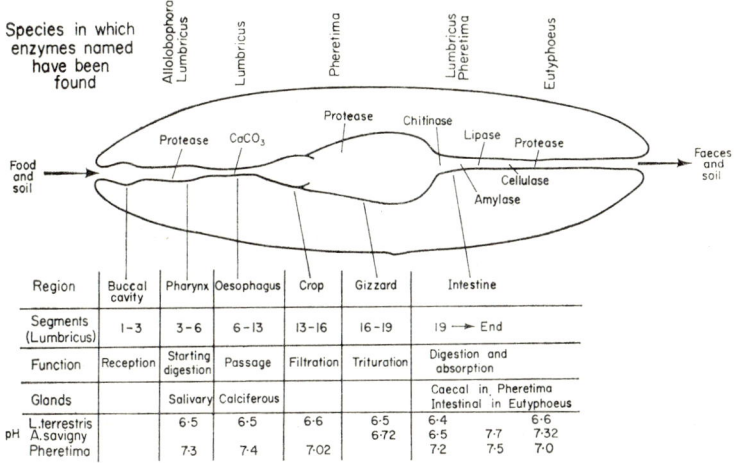

Fig. 7. The alimentary canal, structure and enzyme secretions. The table shows the various regions and the segments in which they are found in *Lumbricus*. The functions of these areas, glands and pH are also indicated.

In view of the catholic eating habits of most oligochaetes, as outlined above, it is not perhaps surprising that in the few species studied there are a wide range of enzymes present in the gut.

Lesser and Taschenberg (1908) made extracts of the gut of *L. terrestris* and incubated them with starch, glycogen, cellulose and inulin. All these substances appeared to be attacked and broken down. Another early study indicated that lichenin, obtained from lichens, is hydrolysed. The enzyme was appropriately named lichenase (Jewell and Lewis, 1918).

Regional localization of the enzyme secretory cells of the gut occurs. Applied closely to the pharyngeal wall at the anterior end of the gut (Fig. 7) is a gland termed "salivary" which secretes a protease in *Allolobophora* and *Lumbricus* (Keilin, 1920). Bahl and Lal (1933) have also described the production of protease by glands in segments 79–83 of the intestine of *Eutyphoeus*, and of amylase by caecae in segments 22–26 of *Pheretima*. Protein is digested in the stomach ( = crop) and anterior intestine of *Pheretima*.

The early work on enzymic digestion, mentioned above, was expanded and elaborated by Robertson (1936) during an investigation of the properties and functioning of the calciferous glands in segments 10–12 of *L. terrestris*. This author was interested in whether or not the calcareous concretions produced in these glands affected the pH of the intestinal contents and consequently the efficiency of the digestive enzymes. He incubated watery extracts of the intestine with, first, starch, and found an amylase activity having an optimum pH of 6·8–7·0, producing an unidentified mixture of sugars as the end product. Secondly the extracts showed the presence of a lipase by the destruction of methyl butyrate at a pH optimum of 6·4–6·6, with the probable occurrence of a second enzyme acting on ethyl butyrate at pH 7·3–7·7. Lastly the extracts were incubated with gelatin or casein for evidence of a protease but little hydrolysis was noted even after incubation for periods of up to 24 hours. Prolonged digestion lasting 48 hours revealed slight activity with an optimum at the alkaline pH 8.

Although Robertson (1936) decided that the proteolytic properties of the intestine of *L. terrestris* are very weak, Millott (1944) was able to show the production of a protease in the anterior intestine (Fig. 8, from Heran, 1956). This enzyme brings about a rennin-like action, clotting milk containing calcium as a co-factor. The secretion of this enzyme is considerably increased by electrical stimulation of the segmental nerves, the result being a diminution of the clotting time of the milk. Rennin in mammals, such as cows and human infants, induces mild hydrolysis of milk but the function of such an enzyme type in coping with the normal diet of an earthworm is not yet clear. It is rather surprising also that Robertson (1936) found little sign of casein hydrolysis when incubated with intestinal extract since casein is the major protein of milk.

Considerable quantities of plant material are included in the food of earthworms, of course, and enzymes capable of utilizing the cellulose of these plants would be of considerable value. As we have seen, Lesser and Taschenberg (1908) suggested that cellulase is present, and this has been confirmed by Tracey (1951). At a pH of 5 the viscosity of solutions of sodium carboxy-methyl-

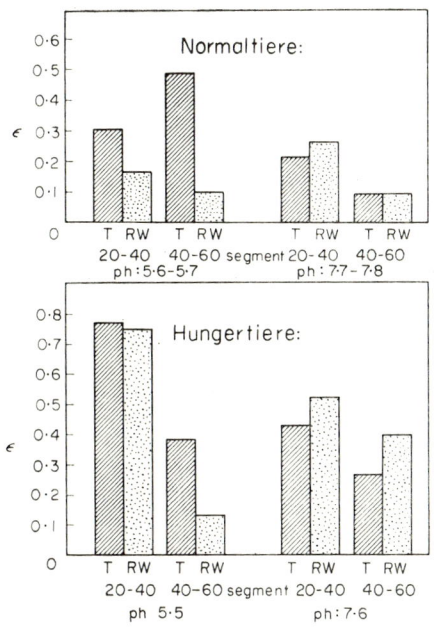

FIG. 8. Protease activity of gut of *L. terrestris*. Above, normal animals, below, starved animals. $\epsilon$ = measure of activity; T = typhlosole; RW = remainder of gut wall in segments 20–40, 40–60. Experiments run at pH indicated (from Heran, 1956).

cellulose is lowered by extracts of the anterior half of the intestine. This is presumptive evidence for the presence of cellulase, and further evidence is provided by the fact that incubation of finely powdered cellulose with gut extracts is followed by the appearance of soluble reducing sugars in the reaction mixture.

Tracey (1951) also found that acetyl glucosamine is produced as an end product of digestion when intestinal extracts are mixed with chitosan hydrochloride at pH 5, indicating chitinase activity. Insufficient material was available to demonstrate the presence of this enzyme in all species but both chitinase and cellulase are found in *A. longa*, *A. caliginosa*, *A. chlorotica*, *Lumbricus rubellus* and *Octolasium cyaneum*, cellulase only being found in *Dendrobaena mammalis* and *Bimastus eisenii*.

The secretion of these enzymes is also a localized affair since very little sign of activity is found in extracts of pharynx, oesophagus, crop or gizzard. The greatest production occurs in the anterior segments of the intestine, only one-tenth of the cellulase and one-third of the chitinase activity of the anterior end being present in the rear portion. It is suggested that these enzymes are actual products of the earthworm gut itself (Tracey, 1951), but in view of the digestion of cellulose by symbiotic bacteria and protozoa in the alimentary canal of other animals such as cows and the shipworm *Teredo* further evidence is required before its final acceptance.

This may, in fact, be partially provided by Parle (1960) who has subjected the intestinal wall to considerable washing before preparing extracts and has found that cellulase and chitinase are still present in fair quantities. Under these conditions it is unlikely that the cellulose decomposers in the intestinal flora and fauna increase to a high enough level to cause appreciable breakdown of cellulose in the time available. The evidence points towards a production of cellulase and chitinase by the earthworm itself, but strictly controlled experiments are still required.

Analysis of worm casts shows that optical rotation of light follows incubation with 20% saccharose at pH 5·5–6·0, due probably to an invertase enzyme originating in the gut. The activity in casts is twice that available naturally in the soil, though it may just as easily be a by-product due to the intestinal flora and fauna which are bound to be associated with cast material (Kiss, 1957).

Each of the enzymes present, protease, amylase, lipase, cellulase and chitinase each have a distinctive optimum pH, but little variation in pH occurs along the length of the gut. The upper and lower limits in *L. terrestris* are pH 6·3 and pH 6·6 respectively as judged by indicator dyes. Electrometric methods of estimating pH have been used on other species e.g. *Allolobophora savigny* (Puytorac and Mauret, 1956) and indicate that the anterior part of the intestine is most acid, pH 6·49, the gizzard being at 6·72 and the rear intestine pH 7·32. The range here is almost a pH unit. Each region of the alimentary canal carries its own distinctive protozoan population. Kagawa (1949) reports pH values of 6·99 for the rear end of the gut and 7·4 in the oesophagus and anterior intestine in *Pheretima communissima* and *P. divergens*.

*Summary*

Oligochaetes are mainly omnivorous, but often selective in their food sources. Enzymes reported in the gut include lichenase, protease, cellulase, chitinase, amylase and lipase. These enzymes break down the components of the food ingested and mobilize it for use by the animal. The occurrence of a cellulase, apparently produced by the intestinal wall of *Lumbricus*, is particularly interesting since many animals feeding on plant material rely on intestinal flora and fauna to hydrolyse cellulose. An invertase enzyme has been mentioned as occurring in worm casts. The enzymes of the gut have distinctive pH optima but the gut contents are remarkably stable with regard to pH since it apparently varies only between about pH 6·3–7·3 along its length.

CHAPTER III

# CALCIFEROUS GLANDS

ALTHOUGH the alimentary canal is virtually a straight tube with but little specialization in its structure, save in the muscular triturating gizzard, there are in most oligochaetes various glands associated with the gut. Some of these were mentioned in the last chapter concerning digestion. The most prominent of these glandular structures, the calciferous glands, were only briefly mentioned and it is here intended to deal with them more fully.

These glands occur as pouch-like diverticula somewhere along the length of the oesophagus, e.g. segments 11 and 12 in *L. terrestris* with an associated oesophageal pouch in segment 10. Within these pouches the epithelium is greatly lamellated or folded. The whole structure is very well provided with blood vessels, the blood flowing through them having come from the absorptive intestine posteriorly. The general gut blood sinus is interposed between the two layers of the lamellae (Fig. 9). A detailed account of the structure of these glands is to be had from Stephenson (1930).

As the name calciferous glands suggests the cells that line these diverticula are glandular in nature and they secrete calcium carbonate. They undergo cycles of activity forming, first, globules in the cytoplasm which, secondly, give rise to spicules of calcium carbonate and these, thirdly, amalgamate to form irregular masses (Bevelander and Nakahara, 1959). These concentrations of calcium carbonate are released into the lumen of the glands and pass from there into the gut. It has, however, recently been suggested (van Gansen, 1959a) that this is not the only method of forming calcareous bodies in the gland. By a method described later (p. 34) the fluid contents of the interior of the gland are concentrated until the calcium carbonate precipitates out *in the lumen* of the gland without ever being in solid form in the cells of the gland.

The elimination of calcium carbonate from these glands has been known at least since 1864 when Lankester suggested that the function of the glands might be associated with the formation of the "egg capsule", but even now, almost one hundred years later, the specific role of the concretions in the economy of the earthworm

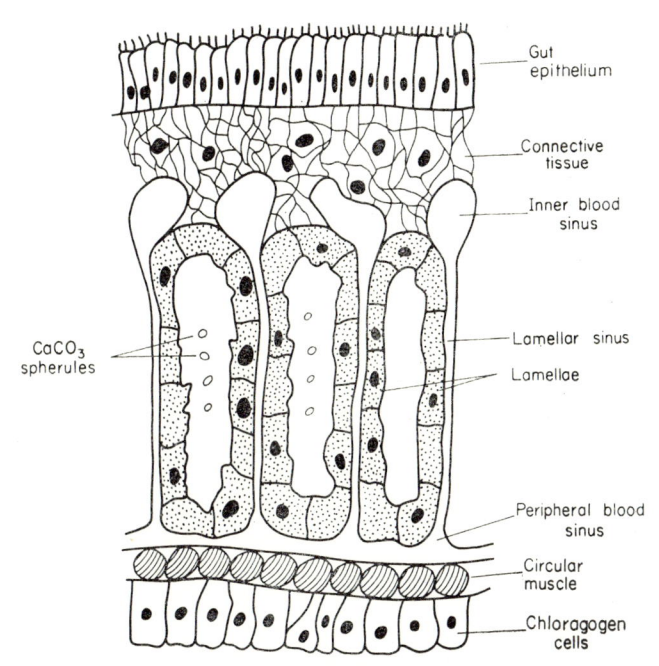

Fig. 9. Transverse section of calciferous glands of earthworm to show disposition of gut epithelium blood sinuses and chloragogen tissue (redrawn from Gabbay, 1958).

is still not completely settled although recent work has made their mechanisms much clearer.

Five major theories appertaining to the function of these glands are mentioned by Stephenson (1930). These are:

(1) The absorption of oxygen.
(2) The excretion of excess calcium (as carbonate) absorbed from the gut.
(3) The absorption of nutritious material from the gut.

(4) The neutralization of the organic acids formed during digestion of vegetable matter contained in the gut; and
(5) The excretion of respiratory carbon dioxide, which we may now read more widely as acid-base balance.

In the early years of this century Combault (1907, 1909) thought the anatomy of the calciferous glands of earthworms closely resemble that of the gills of fish and by postulating muscular movements designed to draw water through the glands he thought that oxygen could be taken up by the extensive blood supply to these structures. Combault noted a difference in the colour of the blood entering the gland and leaving it and thought there was a difference in the oxygen tension of afferent and efferent blood in the glands, but no figures are available to show whether or not this is true. This theory has now generally been discarded. So also has that of Michaelsen (1895).

Michaelsen (1895) considered that the glands are absorptive in function and serve in removing nutrients from the gut contents. The excellent blood supply and lamellated structure provide support for this opinion, but unfortunately the glands lie anterior to the main absorption region, the intestine. As we have seen previously (p. 20), in certain earthworms protein digestion begins under the influence of a protease in the pharynx in some species, but the majority of the digestive enzymes are not encountered in front of the crop and gizzard. Although the glands are well supplied with blood and have a large surface area none the less in many oligochaetes the connection of the glands with the alimentary canal is reduced to a narrow tube, thus rendering the access of gut contents very difficult.

In the two cases considered above experimental evidence has been notably lacking, but the remaining theories have all received a certain amount of support from experimental studies. The ingestion of plant remains from the soil or in some cases, soil alone, very often entails the concurrent intake of organic acids or acid soil particles. The secretion of calcium carbonate has been thought a means of neutralizing this acid content. Robertson (1936) has provided us with a review of various aspects of this problem which has been but little modified in the light of later results. The pH of the alimentary canal of *L. terrestris* lies between pH 6·2 and 6·7 and varies only slightly. Worm casts formed after the passage of soil through the gut have pH values very close to that of the parent soil. Very acid substrates such as moorland peat, lemon juice and other acid media have been observed to cause a decrease in the secretory activity of the gland, supposed by Robinet (1883) and Harrington (1899) to be due to over-production of $CaCO_3$ and subsequent depletion of the gland — but suggested by Robertson, with the aid of X-ray photographs, to be due to a physical disruption of the gland cells by the acid, and as the calcium carbonate concretions can still be identified in casts from very acid media they cannot play a great part in regulating intestinal pH by solution in the gut contents. Rather do the other secretions of the gut, enzymes and mucus, affect the intestinal pH. The optimal range of enzymes for the gut is pH 6·4–7·0 (Robertson, 1936).

The secretions of the calciferous glands also seem unlikely to play much part in any other physical processes of digestion such as crushing and trituration of food. The granules pass unchanged along the gut, and must be of very slight importance compared with the action of soil particles in crushing the food in the gut.

The fourth hypothesis regarding the function of the calciferous glands is due to M'Dowall (1926). This suggests that the calciferous glands excrete excess calcium carbonate absorbed from the diet. Large earthworms are known to favour areas in which limestone and other calcareous rocks are present. In these cases an excess of calcium may be taken in with the diet, but there is no direct evidence that such calcium is in fact absorbed in its passage through the gut. The blood supply to the glands arises from the plexal vessels surrounding the rear intestine, passing anteriorly to the glands. Thus if calcium is indeed absorbed in the intestine the calciferous glands are well placed to remove excessive quantities from the blood, by secreting the concretions into the lumen of the gut. But as they do so at the anterior end of the gut the calcium carbonate aggregations have again to pass along the length of the intestine before reaching the exterior, perhaps indicating that calcium carbonate in the diet is not absorbed, and that calcium uptake is from calcium oxalate and nitrate. It is also not clear that the initial absorption of materials from the gut occurs into the blood vessels. Rather it would seem that the chloragogen cells are the prime movers in absorption, being ideally placed along the intestine, containing many metabolites and the tissue as a whole is associated with many blood vessels, and in this case the calcium absorbed would be eventually passed to the glands but would need to be mobilized first by the chloragogen cells.

Lastly, there is the possibility that the calciferous glands are involved in acid-base balance of the body by a role in the fixation of respiratory carbon dioxide. Robertson (1936) kept earthworms in solutions containing either calcium nitrate or calcium chloride, and after a period of one week analysed the carbon dioxide production in the atmosphere, and also the amount of calcium carbonate present in the soil casts of the animals. As no carbonate was present in solution any carbonate appearing must have come as a result of fixation of carbon dioxide from metabolic sources. The results indicated that less than 10% of the atmospheric $CO_2$ is fixed in the animal body in the form of carbonate voided during casting. A considerable degree of variation occurred, depending on whether or not the animal was active, but an average figure of 5·3% of total carbon dioxide output was produced in the form of carbonate. Consequently Robertson believes that only a small fraction of metabolic carbon dioxide is removed as carbonate. A different approach has been tried by Voigt (1933). He kept earthworms in gaseous atmospheres containing 14% $CO_2$ for 5 days, or in 25% $CO_2$ for 3 days. Under such conditions the contents of the calciferous glands diminish and Voigt believed that in unsuitable conditions of carbon dioxide stress the concretions of the calciferous glands are useful in removing the gas diffusing into the body.

An increase in the rate of secretion of the calciferous glands may be caused by an increase in the atmospheric $CO_2$, but the normal concentration of $CO_2$ is about 0·03%, and this may rise only to 0·25% at a soil depth of 6 inches (Russell, 1950). In wet soil, values far above this may occur for short periods, and as the earthworm in its burrow respires and produces carbon dioxide the concentration may rise even higher.

A direct motor response to high concentrations of gaseous $CO_2$ is reported by Shiraishi (1954). He passed $CO_2$ down tubes so that it impinged on *Eisenia foetida* crawling in the tube. If high concentrations of $CO_2$ (100–25%) are used many animals withdraw their prostomium and anterior extremity and begin to take evasive action. At low gas concentrations (12·5–6%) the reactions are more haphazard and in most cases a momentary stoppage is followed by further progress in the original direction. If it is assumed, therefore, that $CO_2$ may rise to a concentration of 10% in burrows it is unlikely that this will cause the well-known migration which occurs after heavy rain (Darwin, 1881), but it may be reflected by an increased $CO_2$ clearance from the calciferous glands (Voigt, 1933).

Accumulation of carbon dioxide, whether diffusing into the body from a high external concentration, or produced by metabolic reactions within the body, has effects upon the internal pH of the animal. Particularly will this be so in the blood and coelomic fluid, and it has been shown that the oxygen dissociation curve of earthworm haemoglobin is affected by pH (Manwell, 1959).

It is essential, therefore, that some mechanism for the regulation of internal pH be active to counteract changes brought by external conditions or changes in diet constitution. Experiments in which earthworms were kept in 25% $CO_2$ for 16 hours indicate that the pH of coelomic fluid remains virtually unchanged by such treatment. This is taken to indicate that a buffering system is available to maintain a constant internal pH. Removal of the calciferous glands prior to treatment with 5% $CO_2$ for 3 days, however, leads to a depression of coelomic pH, i.e. an increasing acidity, suggesting an accumulation of $CO_2$ in the system. At the end of this period chemical analysis of the tissue shows that heightened levels of calcium are present indicating that calcium is distributed throughout the body instead of being excreted bound to $CO_2$ as calcium carbonate. As a result of these experiments Dotterweich (1933) concluded that the normal activity of the calciferous glands is concerned with regulating the calcium content of the body and that a buffering action is possible in times of carbon dioxide excess. Robertson (1936) supports this view and places further stress on the probable part played by the glands in maintaining acid-base balance of the body.

Further evidence as to the working of these glands has been adduced by Clark (1957). This author investigated distribution of carbonic anhydrase in the tissues of the earthworm. This enzyme is involved specifically in the reversible combination of $CO_2$ with water to form carbonic acid ($H_2CO_3$). In vertebrates this enzyme is found particularly in the red blood corpuscles, where it has a vital role in the transport of $CO_2$ from the tissues to the lungs, in the muscle tissues and pancreas. It is also particularly important in parietal cells of the stomach and in the tubules of the kidney, in both of which situations it is thought to be concerned with the process of hydrogen ion secretion and hence is involved, directly or indirectly, in acid-base reactions.

The situation in the earthworm *L. terrestris* differs from that of the vertebrates in that carbonic anhydrase is completely absent from the blood of the animal, and thus is unable to play a part in the transport of $CO_2$, if such occurs in the blood system. The enzyme is also absent from the coelomic fluid, body wall and the

TABLE 4

CARBONIC ANHYDRASE ACTIVITY OF EARTHWORM TISSUES (From Clark, 1957)

| Tissue | Enzyme units per 50 mg wet wt. |
|---|---|
| Calciferous glands | 80–120 |
| Oesophagus | 30–40 |
| Crop | 12–15 |
| Gizzard | 5–7 |
| Intestine | nil |
| Body wall | ,, |
| Blood | ,, |
| Coelomic fluid | ,, |

intestine. It is, however, present in the gizzard, crop, oesophagus and calciferous glands, increasing in amount in that order (Table 4). We have noted above that carbonic anhydrase is found in systems that cause shifts of pH, as during the secretion of hydrogen ions, by the same token this enzyme has been associated with the solution and deposition of calcium carbonate in other situations, such as during bone formation (Meldrum and Roughton, 1934), and also in the laying down of the calcareous shell of some molluscs (Bevelander, 1952). It thus seems likely that a similar role can be postulated for carbonic anhydrase in the calciferous glands of oligochaetes.

Many organs that have high metabolic rates and that are associated with a secretory or absorptive function, e.g. the distal tubules of mammalian kidney and the planarian nephridium (Danielli and Pantin, 1950), contain considerable quantities of alkaline phosphatase. This enzyme has been demonstrated by histochemical methods in the calciferous glands (Bevelander and Nakahara, 1959). It is associated with mitochondria and is localized at the borders of the cells, the amount present increasing as cellular activity rises, and decreasing as the cell passes into a quiescent phase.

Many mitochondria, of varying types, are found in the cells of the glands (Myot, 1957; Guardabassi, 1957) and electron microscope studies have confirmed the identity of these structures (van Gansen, van der Meersche and Castiaux, 1959).

Two types of cell, however, have been described by van Gansen (1959a) who has provided the most recent work on the glands. This author shows that the glands can be divided into two regions, depending upon the cell type. The posterior portion of the gland, at which the blood arrives directly from the intestine, is composed of cells containing many "bâtonnets" and several types of mitochondria. In the anterior portion of the glands the cells contain large amorphous concretions of calcium carbonate, these not being found in the rear sections. Alkaline phosphatase occurs in both regions, but has different distributions, even in contiguous areas. A small mucus secreting area is present where the glands pour their contents into the gut (Figs. 10 and 11).

The "bâtonnets" of the rearward cells are thought to be infoldings of the cytomembrane of the cells, bearing a close resemblance to the cells of the proximal tubule in mammals (Sjöstrand and Rhodin, 1953). The edges of the cells bordering the lumen of the gland are ciliated and bear small outgrowths reminiscent of a brush border. Cordier (1934) described a very similar series of structures in the wide tube of the nephridium, and Ramsay (1949b) has since shown that this area is the seat of formation of the hypotonic urine excreted by the earthworm, a process probably involving re-absorption of water and salts through the nephridial wall.

Using the similarity in structure of the calciferous cells and those of kidney, nephridial tubules and other actively resorbing systems as a starting point van Gansen (1959a) argues that the two regions of the gland both secrete calcium carbonate, but in different ways.

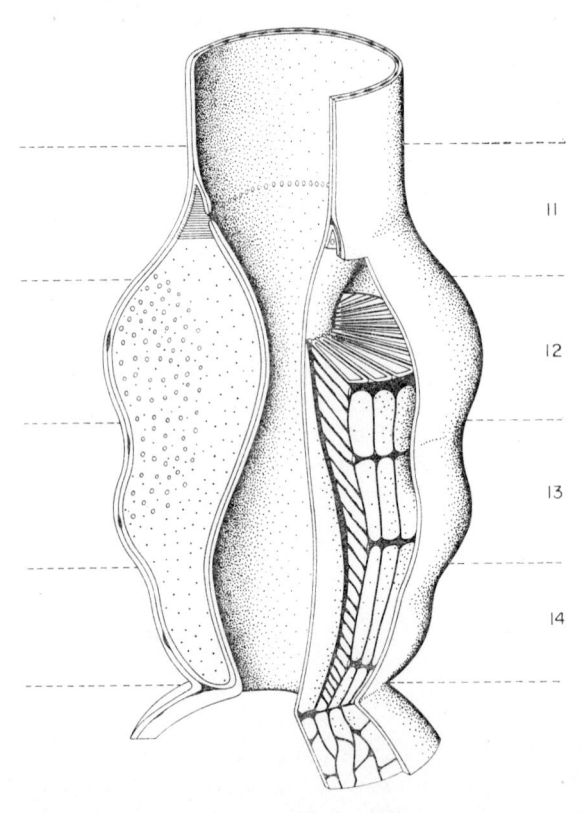

Fig. 10. Anatomy of calciferous glands of *Eisenia foetida*. To the right a vertical section passing outside the chambers of the gland to show the blood sinus system. In segment 12 a transverse section of the chambers to show distribution. To the left a vertical section passing through a chamber. Sinuses marked in black, calciferous zones with small circles, and mucus areas with horizontal hatching. 11, 12, 13, 14 indicate numbers of segments (from van Gansen, 1959a).

The posterior region receives its blood supply direct from the intestine and the fluid contents of the gland cells are almost identical with the blood. By a process of filtration, which presumably removes protein and other high molecular weight compounds, the fluid enters the lumen of the gland. The cells then actively take up water and certain salts via the brush border and

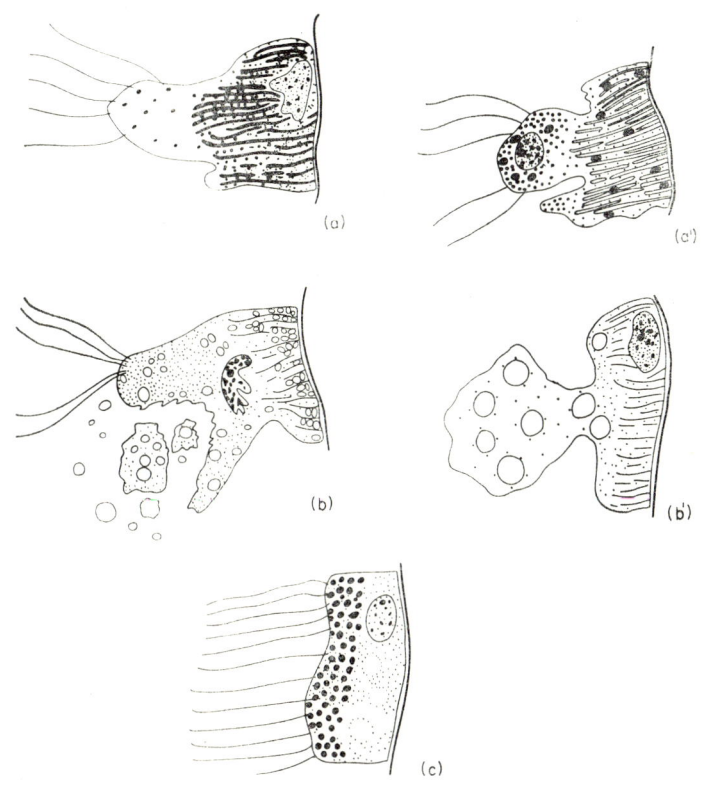

Fig. 11. Cytology of calciferous cells. (a) (a′) from posterior region. Altmann–Fuchsin stain showing bâtonnets and several types of mitochondria. Alkaline phosphatases (black), esterases (hatched areas). (b) (b′) cells of calciferous area with spheroliths abundant in the cells. (c) Mucus cells. Black circles show polysaccharide inclusions (from van Gansen, 1959).

alkaline phosphatase system. As a result the contents of the lumen are concentrated and calcium carbonate eventually precipitates out (Fig. 12).

By the time the blood reaches the mid and anterior portions of the gland its constitution has been somewhat altered. The calcareous aggregations form within the cells as a consequence and are released by rupture of the cells into the lumen.

It would seem, therefore, that the calciferous glands are an organ of excretion in the earthworm, effective in salt and water balance, and presumably assisting the nephridia to maintain a constant internal composition by removal of calcium carbonate. This would be of particular value to species living in calcareous habitats. Indeed such species are known to have much larger

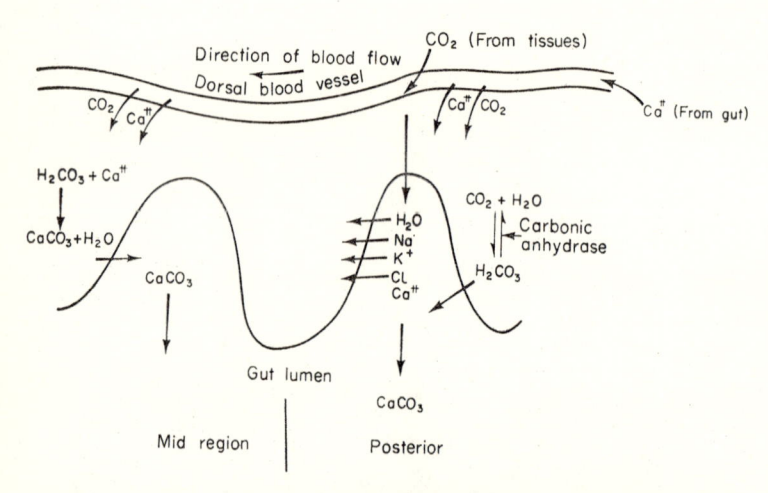

Fig. 12. Formation and secretion of calcium carbonate concretions from two areas of the calciferous glands (after van Gansen).

organs than species from less calcareous soils. And its probable role in the removal of carbon dioxide and maintenance of a steady internal pH must not be forgotten.

*Summary*

The calciferous glands of earthworms secrete calcium carbonate concretions into the lumen of the gut. Blood flows forward from the intestine to these glands, the concretions are passed into the gut and move posteriorly to be voided unchanged through the anus.

Various theories as to the significance of these glands have been put forward from time to time. It now seems reasonably settled that these glands are organs controlling the acid-base balance of the body. They fix a certain percentage of the metabolic carbon dioxide produced by the body, and this can be changed according to the amount of atmospheric carbon dioxide present. Coelomic pH is maintained at a stable level, but removal of the calciferous glands is followed by an increasing acidity of the coelomic fluid. Carbonic anhydrase, an enzyme concerned in acid-base reactions in other animal tissues, is present in large amounts in the glands. An outline is presented of the mechanisms by which calcium carbonate is secreted by the glands, two processes occurring dependent upon which section of the gland is considered.

CHAPTER IV

# THE AXIAL FIELD

THE metamerically segmented body of the earthworm, with its serially repeated organs occurring along the length of the body, and with a certain degree of specialization at the anterior end, has been a favourite choice for investigators concerned with activity gradients, or fields, in bodily functions. From analysis of the reactions of the tissues to poisons, the electrical properties of the body, respiratory rates, or even chemical content, it has many times been postulated that an axial field of dominance is exhibited in oligochaetes, the anterior end exerting a controlling effect over the rest of the body. This is of particular importance when considering the phenomenon of regeneration, a process believed to be governed by the disturbance of a static axial gradient. Even now, however, there is disagreement as to the existence of such gradients, let alone the influence of such states.

In 1916 Hyman investigated the effects of cyanide upon several species of oligochaetes including *Dero*, *Lumbriculus*, *Tubifex* and *Limnodrilus*. She found that the susceptibility of the tissues of these worms was not constant and that death occurred most quickly at the anterior end, the more posterior segments resisting cyanide poisoning relatively longer (Fig. 13). At the very rear end the segments were rather more sensitive to cyanide than in the mid-region of the body and consequently a U-shaped curve was obtained, with the most sensitive regions at front and rear, and the most resistant region in the middle of the length. On this basis Hyman suggested that the more rapidly metabolizing portions of the body were the most sensitive.

Shearer (1924) therefore undertook to study the respiration rates of pieces of earthworm taken from various regions along the length of the body. Using a Haldane gas analysis apparatus he examined both live segments and acetone powder extracts from the same areas, and from his results concluded that the anterior end respired faster than the posterior end. He postulated further that the anterior end possesses greater organization and a definite substance is produced here that diffuses back through the body to affect the respiration of the more rearward segments, the further from the head the lower the respiration rate since less material

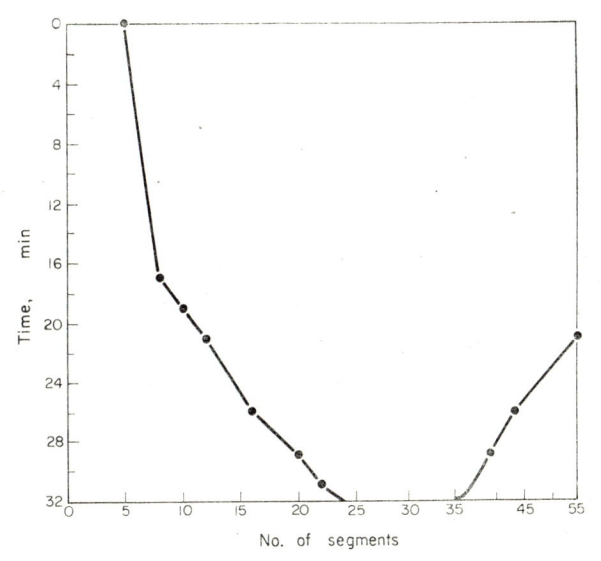

Fig. 13. Graph of the axial gradient of *Dero limosa* showing secondary posterior rise. Based on time taken for disintegration to occur when treated with cyanide (from Hyman, 1916).

would be reaching these segments. He did not observe a further rise in respiration at the extreme rear end.

Further evidence for the dominance of head over tail was adduced by Perkins (1929). He repeated the work of Shearer on respiratory rates and obtained the highest values from pieces of the anterior end. The rate decreased steadily the further the segments were from the head to about 60% of the length of the body and then rose again slightly towards the rear extremity, thus mimicking the curve of Hyman for poison susceptibility. Other parameters also apparently showed a graded rise and fall over body length. The extractable sulphydryl and sulphur content and iodine equivalence values all showed a peak concentration at about the point where a divided worm grows either forward or backward depending upon the aspect of the cut surface. The concentrations fall on either side of this point and the curve does not resemble that of oxygen uptake. Unfortunately Perkins's graph has no numerical information and there is no indication as to the methods used.

Criticism of these results and others similarly inclined upon such groups as the planarians, led Shearer (1930) to re-investigate the respiratory uptake using a manometric technique, more delicate than his previous method. As a result of this work he withdrew his previous conclusions completely and suggested that the apparent differences between anterior and posterior pieces are due to differences in motility, i.e. the slices from the anterior end tend to move around the reaction vessels more rapidly than the posterior pieces, the anterior fragments therefore metabolize quicker and consequently the oxygen uptake is faster. He noted by eye that the muscular activity of anterior slices was more pronounced, but he used rather larger fragments than is advisable in Warburg methods.

Much smaller pieces were used by Maloeuf (1936b). He isolated fragments (20–30 mg) into Frog Ringer and used the Warburg method to obtain the respiratory rate at 30 °C, a temperature higher than anything liable to be encountered by *Lumbricus* in the course of its daily life in temperate regions, and shown later to be above the thermal death point for *L. terrestris*, (28 °C) (Hogben and Kirk, 1944; Wolf, 1940). His results showed that there were no consistent differences in respiration of slices taken from different regions, except that the anterior end respired more *slowly* than the posterior end. The size of the pieces used was small enough to eliminate co-ordinated movement and thus all the respiratory rates were reduced to similar values, a fact also conspired by the high temperature which would tend to destroy the metabolic activity of the pieces. Watanabe (1927) described U-shaped gradients for carbon dioxide production, oxidizable substances, electrical potentials and total solid content along the longitudinal axis of the body wall of *Pheretima hilgendorfi* and *Eisenia foetida*, and in *Stylochus* Watanabe and Child (1933) demonstrated a similar curve for $CO_2$ output, indophenol oxidase reduction and KCN susceptibility along the length of the animal. Strelin (1938), however, was unable to obtain similar results using dye techniques on *Limnodrilus* and *Tubifex*.

It is obvious that if a gradient or polarity exists between the anterior and posterior ends it may have considerable importance in the process of regeneration when the normal picture is disturbed and several workers have concerned themselves with this aspect of the problem.

The freshly cut end of an earthworm is generally electronegative with respect to the neighbouring uninjured segments (Morgan and Dimon, 1904) and this observation was extended to cover the body by Watanabe (1927). In *E. foetida* the anterior end is electronegative with respect to the posterior end on the dorsal surface. Ventrally the negativity is greatest at the anterior end, falls in the mid-region (segments 65–70) and then rises again towards the rear end. In fact the curve is U-shaped.

This variation in the electrical potential was confirmed by Moment (1949) also working on *E. foetida*, and he noted that the potential decreased from $+15$ mV to $-11\cdot8$ mV immediately after the worm was divided at the 50–51 segment furrow. As regeneration of the rear end proceeded so the potential steadily returned to its original level, reaching this point after approximately 3 weeks. After this time, although the animal continues to lengthen no further segments are added and no change in potential is seen.

Kurtz and Schrank (1955) repeated this work and obtained essentially the same results. The anterior end is 14 mV negative with regard to the hind end, and the clitellar area has greater negativity than the same area in non-clitellate worms (Fig. 14). Upon starvation the observed voltage decreased posteriorly and at the clitellum. This was particularly marked in the first 4 days of starvation, but a slow constant decrease was recorded up to 30 days starvation. These workers, however, did not find a reversal of polarity directly after severing the body at the 50–51 segment region, and the regeneration process was accompanied by fluctuating voltages rather than by a steady rise. When the regenerant is fully grown the anterior–posterior potential may even show a rise into positive values.

D

Both Moment (1949) and Kurtz and Schrank (1955) consider that the results outlined above support a thesis that the formation of new segments in response to a loss of part of the body is correlated with the electrical potential of the intact body. The average number of segments regenerated by *E. foetida* lies between thirty-three (Kurtz and Schrank, 1955) and forty (Moment, 1949), and

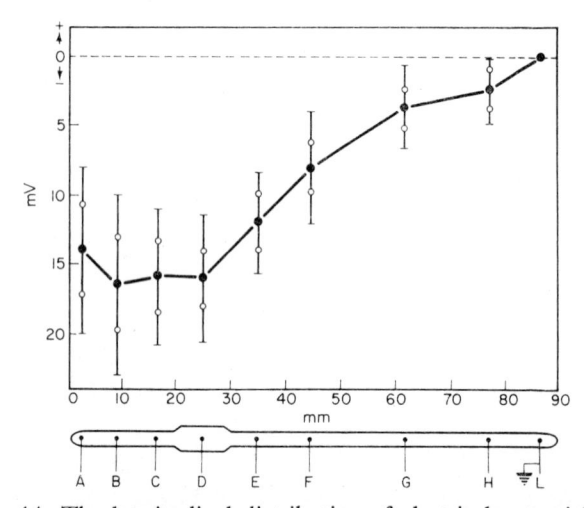

Fig. 14. The longitudinal distribution of electrical potentials of earthworms measured on the dorsal side with the posterior end earthed. Each point represents an average of 200 measurements and standard deviations are shown by the vertical lines. Small circles on the vertical lines show the reliability of the means at 5% level of confidence. The regions are located thus: A, segments 2–4; B, segments 8–12; C, 20–23: D, 26–32; E, 35–38; F, 1 cm. posterior to E; G midway between F and H; H, 1 cm. anterior to L; hindmost 4–5 mm (from Kurtz and Schrank, 1955. Copyright 1955, University of Chicago).

the final voltage difference between head and tail reaches 20 mV (Kurtz and Schrank, 1955). These authors suggest that when fully grown a critical voltage has developed between the extremities and this inhibits further prolongation by the addition of fresh segments. It would be interesting to know what the voltages are in young growing worms and whether a similar voltage difference exists from hatching onwards. It would also be interesting to know why growth does not stop at the clitellar region where the voltage is at its lowest and the overall potential difference at its greatest. The results of Moment, and Kurtz and Schrank show discrepancies for the same species, and no explanation is forthcoming as to how a critical inhibitory voltage may affect metabolism in the cells, nor indeed whether the voltage changes are the prime mover or the end result of regeneration. One hopes that some of these points may be settled in the future, especially in view of the work of Hubl on neurosecretory phenomena in regeneration.

The question of metabolic gradient has been re-investigated most recently by O'Brien (1947, 1957a). He examined a number of systems in *A. longa* using modern techniques.

TABLE 5

RESPIRATORY RATES OF VARIOUS SEGMENTS OF THE BODY OF *A. longa* ($\mu$l./100 mg wet wt.) (From O'Brien, 1957a)

| Segs | 11–20 | 21–35 | 35–55 | 55–70 | last 10 |
|------|-------|-------|-------|-------|---------|
| Rate | 11·4  | 12·7  | 12·0  | 11·9  | 17·2    |

The last ten segments of the body respire at a rate faster than that of preceding segments. As O'Brien used the Warburg technique and his slice thickness was only 0·6 mm the objection of Shearer (1930) and Maloeuf (1936) regarding possible muscular activity is unlikely to be critical, but there are no pronounced differences in his figures except that the rear end respires faster than the other regions (Table 5). Removal of the hind end was followed by changes in respiratory rate. The oxygen uptake of the stump drops, but as the regenerant grows the respiratory rate of the hind end (new tissue) rises (Table 6). Segments in front of the stump remain at a normal rate of respiration. The higher values obtained for the rear end are interesting in view of Needham's suggestion (1957) that there may be a posterior metabolic and hormonal integrating centre.

O'Brien (1957a) also subjected the body of two species, *E. foetida* and *O. cyaneum*, to analysis for various metabolic substances. He found, for example, that the distribution of glycogen in the muscular body wall (average 4·5 mg/g wet weight) which was low in the region of segments 36–55 was the mirror image of the distribution of glycogen in the intestinal tissue (average 7·0 mg/g wet wt.) which had an extended peak in the same area. Lipoid materials (3·4 mg/100 mg wet wt. in the whole worm) show a maximum in the middle part of the body, probably due to the intracellular absorption of lipoid into the intestinal epithelium and eleocytes (Fig. 15).

TABLE 6

EFFECT OF REGENERATION ON THE RESPIRATORY RATES OF AREAS OF THE BODY OF *A. longa* ($\mu$l./100 mg wet wt.) (From O'Brien, 1947)

| Time hr | Regenerant | Tissue adjacent | Mid-region |
|---------|------------|-----------------|------------|
| 48      | 8·2        | 8·5             | 12·9       |
| 72      | 10·1       | 7·1             | 13·1       |
| 168     | 11·4       | 8·3             | 12·6       |

The production of lactic acid, reflecting the glycolytic activity of the tissues, shows a U-shaped curve, and succinoxidase activity of body wall and granules prepared by centrifugation also show greater values in preparations from anterior and posterior regions than in mid-body.

These later experiments involved homogenates and not continuous tissues so it is improbable that muscular movement can account for regional differences. Unfortunately once again the temperature at which these experiments were run was near the upper thermal death point, 27 °C, and this may have had some effect. The greater concentration of functional organ systems at the anterior end, in the shape of gut specializations, sex organs and nervous concentrations may all serve to increase metabolic reactions in this region, whilst the mid-portions of the body, containing the unspecialized intestine and nephridia might be expected to have a lower metabolic rate, but apart from Needham's idea of a posterior penultimate control centre no one has yet suggested the

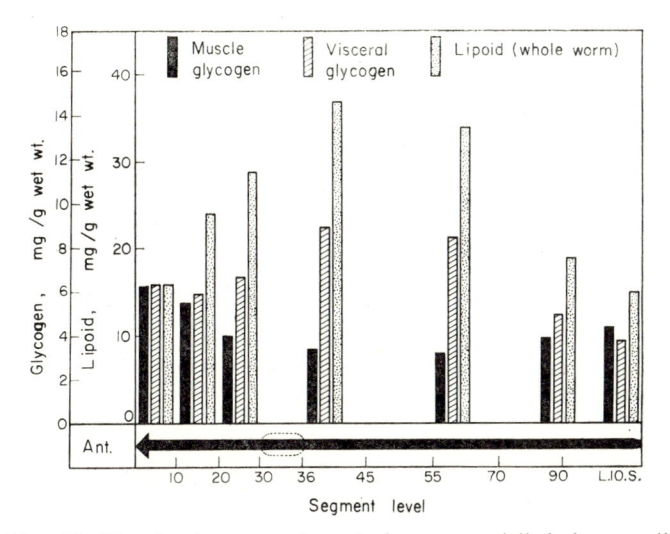

FIG. 15. Muscle glycogen, visceral glycogen and "whole worm" lipoid related to segment level. Clitellar zone extends from segments 29 to 36. Each zone analysed consists of approximately 10 body segments. The posterior zone contains the last ten segments (usually 100–110). Histograms represent mean of five separate analyses; for each analysis large non-clitellate specimens of *E. foetida* were used. Glycogen estimations thirty worms per lot. Lipoid estimations twenty worms per lot. Results expressed as mg/g wet weight of worm tissue (from O'Brien, 1957a).

cause for heightened activity at the rear extremity. It is generally true that in pigmented species the hind end contains more colouration than the mid-regions and this may be correlated with light sensitivity, particularly useful in species casting on the surface, such as *A. longa*.

*Summary*
Axial gradients have been reported in many of the various

components of the oligochaete body. U-shaped curves have been obtained for the susceptibility to poisons, respiratory rates, sulphur content, solid content, carbon dioxide production, lactic acid, oxidizable substances and electrical potential. Influence on the rate of regeneration has been suggested as a function of the electrical field of oligochaetes.

CHAPTER V

# NITROGENOUS EXCRETION

THE DIET of any animal contains a multitude of organic compounds, carbohydrates, fats, amines and proteins among them. From these the animal builds its own structure and obtains energy for the many enzyme systems necessary to it. The protein of the cells and secretions is manufactured from the amines formed by hydrolysis and degradation of the proteins in the food. These are absorbed, transported to the sites utilizing them and the characteristic proteins of the body are then formed. This is not the end of the story, however, for as was shown by the classic studies of Schoenheimer and Rittenberg the body proteins are in a continual state of flux, changing constantly. The materials no longer required are removed, broken down to non-toxic units and excreted. In some cases they may be excreted little altered e.g. creatinine from creatine in vertebrates, but in the majority of cases the nitrogen fragment is converted to a non-toxic substance and removed in the urine. If the animal has access to voluminous quantities of water ammonia is the most common and most easily voided excretory product e.g. in amphibians, fish and marine invertebrates. In land-dwelling species water conservation is of great importance and the production of small quantities of urine containing high concentrations of nitrogenous materials is the rule. The less toxic products urea and uric acid make their appearance in these forms.

Nitrogen is also lost in many cases in the form of muco-proteins, secreted by epidermal cells on to the surface of the body for the purposes of lubrication, maintaining a clean body surface and providing a buffer against the external environment. Oligochaetes are no exception to these generalizations, and once again the vast majority of the published work has been concerned with earthworms.

One of the earliest attempts to ascertain the excretory products of earthworms was made by Lesser (1908) on *L. terrestris* and *E. foetida*. Using the limited techniques available in those days he found no evidence for the occurrence of uric acid, urea or allantoin, but did obtain results indicating the presence of ammonia. Much later Florkin and Duchateau (1943) showed that this was not altogether surprising for earthworms possess no allantoinase, allantoicase or uricase and cannot hydrolyse allantoin, allantoic acid or uric acid respectively. These authors did find that xanthine oxidase, the uric acid forming enzyme, is apparently available. Urease has been reported by Przlecki (1923) to break down urea and form ammonia at the very slow rate of 2·28 mg ammonia/100 g/day. This may have been erroneous, however, and perhaps due to bacterial action for Abdel-Fattah (1957) found no signs of urease activity in blood, coelomic fluid, body wall or gut of *L. terrestris* or *A. longa*, although ammonia and urea were both to be found in these tissues.

The nitrogenous excretion of earthworms may be considered as occurring in two fractions, each accounting for approximately equal portions of nitrogen removed each day. The first portion consists of a protein amounting to about 0·030 g/100 ml urine/day (Bahl, 1947a, b), and is most probably derived from the mucus secreted copiously by the body wall. This mucus acts as a lubricant as the worm proceeds along its burrow, also helping to bind soil particles together and preventing the burrow walls from collapsing. Mucus also probably acts as a buffer system outside the body since it is secreted in large amounts when the animal is immersed in a noxious stimulant such as acid. This mucoid protein can account for about half the total nitrogen lost each day (Needham, 1957; Haggag and El-Duweini, 1959), though this may be a reflection of the somewhat unnatural conditions encountered by the animals in experimental vessels.

The second fraction, representing the end products of metabolism, is a fluid urine comprising a mixture of ammonia and urea, with rather confused reports also on the presence of uric acid and allantoin (Fig. 16). The proportions and presence of these nitrogenous substances depends upon the species and upon the condition of the animals when examined.

Lesser (1908) and Delaunay (1934) both isolated earthworms

into large volumes of water for varying periods of time, and then analysed the resulting fluid. As we have mentioned above Lesser had difficulty in deciding what nitrogenous substances are present, being certain only of ammonia. Delaunay (1934) found that ammonia is the largest single excretory product detectable in the urine, and this is allied to urea, and a little amino nitrogen. He found no uric acid. Under the conditions of the experiments, however, considerable dilution of the urine occurred and no reliable

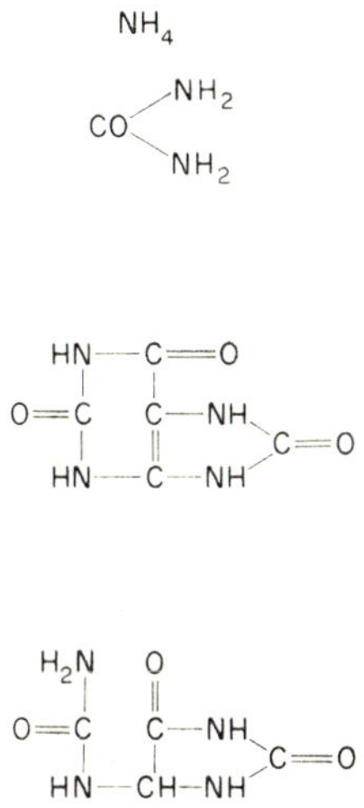

$$NH_4$$

$$CO\genfrac{}{}{0pt}{}{NH_2}{NH_2}$$

Fig. 16. Formulae of ammonia, urea, uric acid and allantoin, the presence of which have been reported in the urine of earthworms.

estimate can be made as to the volume and chemical concentration of the urine produced. More recently results on urine composition have been obtained from earthworms placed in small volumes of water (Needham, 1957), in clean flasks washed out periodically (Cohen and Lewis, 1949b), in moist air with urine allowed to drain away down a slope (Bahl, 1947a) and by collection of urine direct from the nephridiopores (Ramsay, 1949a). Unfortunately the last-named method, by far the most accurate technique, has been used so far only for estimating the osmotic pressure and mineral salt content of the urine, and the concentration of nitrogenous compounds has not been determined in the small quantities available by this micropipette method.

The losses to evaporation and dilution by excess fluid are, however, minimal in the experiments of Cohen and Lewis (1949 a, b) and Needham (1957) but somewhat conflicting results have been reported by these workers. Normal, healthy, fed *L. terrestris* excrete an average of 5·75 μg $NH_3$/g wet wt./day, which represents approximately 70–80% of the total non-protein nitrogen. Urea accounts for about 8% of the total (0·37 μg/g wet wt./day): Cohen and Lewis (1949 a, b) record 0·16 μg/g wet wt./day allantoin and 0·113 μg/g wet wt./day uric acid, substances not found by some other authors. A period of starvation is followed by an inversion in excretion of the major constituents, ammonia falling to 1·8 μg/g wet wt./day, whilst the urea fraction rises to 18·15 μg/g wet wt./day. At the same time the total amount of non-protein N excreted rose from 7·4 μg/g wet wt./day to 21·5 μg/g wet wt./day.

Very different absolute values were obtained, however, for the same species, by Needham (1957). Estimates of the total non-protein nitrogen content of urine in animals feeding on elm leaves amounts to 94·8 μg/g wet wt./day and starvation raises this level to 159·2 μg/g wet wt./day. The major components of nitrogenous excretion are confirmed as ammonia (54 μg $NH_3$/g wet wt./day) and urea (40·8 μg/g wet wt./day). Starvation of the animals alters the relative importance of these two substances, the output of urea rising by some 300% whilst ammonia decreases 30%. Thus although ammonia is present in excess of urea in normal feeding earthworms and the changes seen upon starvation are in the same direction as those noted by Cohen and Lewis (1949b), they differ considerably in the ratios of $NH_3$ : urea, and in absolute values by a factor of approximately 10 (Table 7).

In his 1957 paper Needham reports on observations made on two species of earthworms, *L. terrestris* and *E. foetida*. These two species are fairly closely related but show dissimilar patterns of nitrogen excretion. Table 7 compares the nitrogen output of these two species and reveals that whilst the urea production of the two forms is closely similar there is a considerable discrepancy between the ammonia released by each species, that of *E. foetida* being double that of the much larger *L. terrestris*. Needham finds no evidence that the roles of urea and ammonia are interchanged in the

TABLE 7

LEVELS OF EXCRETORY SUBSTANCES REPRESENTED IN THE URINE OF VARIOUS EARTHWORMS

| Species | L. terrestris B | L. terrestris A | L. terrestris B | L. terrestris A | E. foetida B | E. foetida A | A. caliginosa B | A. caliginosa A | Pheretima |
|---|---|---|---|---|---|---|---|---|---|
| Urea | 0·37 | 18·15 | 40·8 | 129·1 | 48·3 | 73·1 | 19·9 | 83·2 | 3·24 |
| NH₃ | 5·75 | 1·85 | 54 | 30·1 | 104·85 | 228·8 | | | 2·66 mg/100 ml |
| Amino N | | | | | 51·7 | 43·6 | 37·4 | 14·0 | |
| Allantoin | 0·16 | | | | | | | | |
| Uric acid | 0·113 | | 1·42 | | | | | | |
| Total non-protein N | 7·4 | 21·5 | 94·8 | 159·2 | 204·8 | 345·5 | 87·5 | 133 | |
| Author | Cohen and Lewis, 1949 | | Needham, 1957 | | Needham, 1957 | | Haggag and El-Duweini, 1955 | | Bahl, 1947b |

B = Before starvation
A = After starvation in µg/g/day

excretory picture of *E. foetida* after a period of starvation, both bear the same relationship to one another although the actual quantities involved are doubled. *L. terrestris* on the other hand shows a trebled output of urea after inanition, but a slight drop in the ammonia production. *A. caliginosa*, another small species, shows a nitrogen excretion of similar pattern to *L. terrestris* though it does not excrete such large quantities of urea or ammonia.

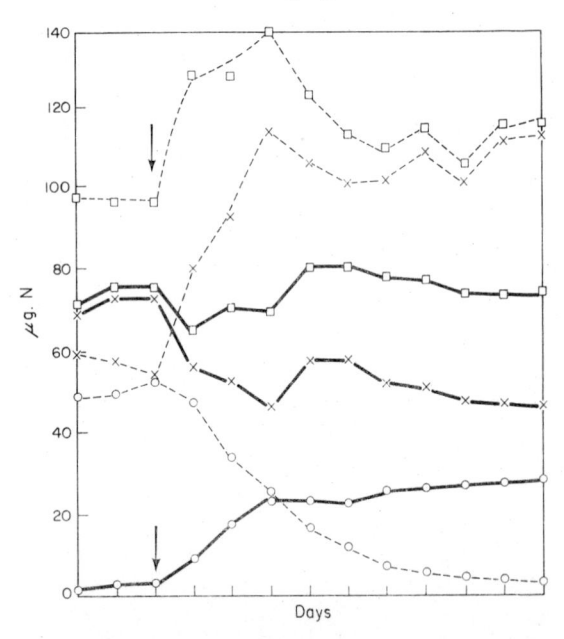

Fig. 17. Excretion of urea N(×) amino+ammonia-N(○) and amino+ammonia+urea-N (□) *Lumbricus* during transition from feeding to fasting (- - - - -) and from fasting to feeding on elm leaf (———) in μg/g body wt./day (from Needham, 1957).

Fig. 17 shows the course of excretion over a period of days in *Lumbricus*.

Uric acid accounts for only a very small percentage (1·5%) of the total nitrogen excreted in fed *L. terrestris* and a rather larger proportion in *E. foetida* (3·4% in fed animals, 3·35% in starved individuals). In this work no mention is made of allantoin. When the posterior region of the animal is removed the excised portion excretes nitrogen at a high rate, mainly as protein, until the piece dies.

Urine as excreted from the earthworm body is acidic. The acidity is shown to vary from time to time and starving animals produce less titratable acid than do feeding worms. *E. foetida* produces more acid than does *L. terrestris*, but whereas the pH of the urine is buffered by urea in the case of *L. terrestris*, as shown by a rise in pH upon treatment with urease, this is not so in *E. foetida*. Needham (1957) suggests that in the latter case the buffering of the urine is a function of some non-nitrogenous akali, citing as a possible mechanism the calcium carbonate removed from the calciferous glands.

Thus in these two related species the excretory products differ in a number of ways. *E. foetida* has a much higher specific output of total nitrogen and of the ammonia fraction than does *L. terrestris* and ammonia makes up a larger part of the total excreted. The ammonia excreted does not go down, but rather goes up, when the animal is subjected to fasting. *E. foetida* has a higher removal of acid, and it is believed utilizes a non-nitrogenous base to neutralize it, possibly the calcium carbonate from the calciferous glands, which is thought to be playing a part in the internal regulation of pH as well (Voigt, 1933). In the conditions of Needham's experiments the anal and nephridial excreta were able to mix freely, and thus his explanation might be true for this set of circumstances, but in the field the animal is free ranging in a burrow, and as it is probable that urine from the nephridia is immediately absorbed on to soil particles as it is released, it seems unlikely that concretions issuing from a distant site will have much effect upon the urine pH.

Cohen and Lewis (1949b) and Needham (1957) provide clear demonstrations as to variations in excretory products of earthworms when they are starved, although they do not agree on the amounts involved. That the geographical location may be important in arriving at results and conclusions seems to be indicated by the work of Haggag and El-Duweini (1959). *A. caliginosa* excretes nitrogen in a pattern similar to that of *L. terrestris* (Needham, 1957) though the absolute amounts are much smaller. In Egyptian representatives of this species protein accounts for up to 50% of the total nitrogen excreted, similar to that reported by Needham.

Unfortunately Haggag and El-Duweini (1959) do not give details of the nitrogenous content of urine in conditions of no starvation, but figures for excretory values for periods of up to 27 days without food reveal that no systematic changes in the proportion of ammonia or urea occur. In animals starved for 5 days the ammonia fraction accounts for 54·2% of the total excreted nitrogen, whilst urea amounts to 6·25% and uric acid 0·26%. No indication, however, is given of the actual quantities involved. Prolonged inanition leads to alterations in the amounts of these substances represented in the urine but in no case does the urea content exceed that of ammonia. This is in direct contrast to the results of Needham (1957) on the same species in England.

The availability of water in the environment may obviously vary considerably from time to time and such variations may be of great importance in the excretory mechanisms utilized by oligochaetes. The change to a terrestrial habitat made by many oligochaetes means that water is always at a premium. We shall see later that the earthworm is rather like a fresh-water animal in its water relationships. In animals living in a great excess of water the more highly soluble nitrogenous products such as ammonia and trimethylamine oxide are very often the sole excretory products, but with the changeover to the land and a decreasing volume of available water first urea and then uric acid becomes the dominant substance, each less toxic than the preceding one.

The earthworm hasn't settled upon one excretory substance to do the job, relying mainly on ammonia, which is removed into the soil moisture usually found in excess around the body, but also excreting urea. The amount of the latter that is excreted seems to depend upon the metabolic water available since starvation leads to a decrease in ammonia and an increase in urea content of the urine. Needham (1957) immersed earthworms in varying quantities of water, for varying periods of time, and concluded that the volume of water was unimportant unless accumulation of toxic products was allowed to occur over a long period. Again this is unlikely to occur in the field. No experiments have so far been carried out on the effect of dehydration upon the excretion of these animals, a state of affairs just as likely to happen in nature, but obviously involving technical difficulties in the collection of adequate samples for analysis. The nephridiopore collection method of

Ramsay (1949 a, b) might have a useful application here though the quantities obtained would be extremely small.

Although no information can be presented regarding the changes of urine content, if any, during desiccation it is possible to show that the tissue content of nitrogenous substances changes its nature during such conditions. Analysis of tissue homogenates from fresh *A. caliginosa* show that ammonia nitrogen amounts to 26·48 mg/g tissue nitrogen, urea 13·14 mg/g N and uric acid 1·33 mg/g N. These values, like those for urine content, do not change during starvation (in the experiments of Haggag and El-Duweini, 1959). When the animals are allowed to dry in air, however, the ammonia and uric acid content remain quite stable but urea values rise considerably (Table 7).

This rise in urea content can be correlated with a gradual decrease in the amount of water present suggesting that the earthworm is able to adapt its excretory pattern in times of internal water stress by carrying the metabolic cycle further to produce urea instead of ammonia. The production of the even less toxic uric acid, however, is not affected, perhaps due to a limited enzymatic apparatus (xanthine oxidase) which is unable to deal with the increased conversion required (Florkin and Duchateau, 1943).

Haggag and El-Duweini (1959) made no differentiation between the various organs of the body, homogenizing the animals entire without previous separation of the parts. That there is a partition of ammonia and urea between the gut and the body wall has been shown by Heidermanns (1937) for *L. terrestris* and by Bahl (1947b) for *Pheretima posthuma*. Analysis shows that the gut wall contains more urea than the body musculature, and also more ammonia, though the differences here are not so pronounced (Table 8).

Creatinine was reported by Bahl (1947a) as a breakdown product in both body wall musculature and intestine in *P. posthuma*, and Abdel-Fattah (1957) obtained a figure of 1·60 mg% for the creatinine content of urine from *Lumbricus* and *Allolobophora*. The methods used in these estimations are not, however, specific and with the discovery of lombricine in oligochaetes, and other guanidine derivatives in sundry other annelids it is highly likely that the so-called creatinine fraction is in fact some other guanidine product.

From these analyses it is apparent that urea and ammonia occur as metabolic products in the tissues, notably the intestine, passing from there into the blood or coelom and from thence to the exterior via the nephridia. The ability of the intestinal wall to form urea was first shown by Heidermanns (1937) who incubated homogenates of this organ with "peptone" and obtained urea, this substance not being formed when homogenates of the body wall were incubated with the same substrate. The provision of arginine or ornithine as substrate had no effect upon the elaboration of urea and it was concluded that urea is not formed by the Krebs–Henseleit series of reactions.

TABLE 8

NITROGENOUS CONTENT OF TISSUES OF TWO EARTHWORMS
(values mg%)

|  | P. posthuma gut | P. posthuma body | L. terrestris gut | L. terrestris body |
|---|---|---|---|---|
| NH$_3$ | 4·3 | 3·6 | 7·7 | 3·3 |
| Urea | 4·6 | 2·1 | 15·3 | 1·6 |
| Creatinine | 4·4 | 5·5 |  |  |
| Author | Bahl, 1947b |  | Heidermanns, 1937 |  |

Heidermanns used simple water extracts in his work, and repeating his observations Cohen and Lewis (1950) were also unable to obtain a yield of urea when intestinal preparations of *L. terrestris* were incubated with arginine. They were also unable to find urea after incubation with "peptone", a result in direct contrast to that of Heidermanns, and for which they were unable to account. However, using tissue homogenates in phosphate buffer solutions, arginase activity is found (Cohen and Lewis, 1950) when the intestine alone is considered, but if the whole body is macerated and incubated then no arginase is observed. The optimal pH for the enzyme is pH 7·2–8·6, and a co-factor in the form of $10^{-3}$M cobalt increases the activity. Fresh captured and recently fed animals are found to possess only slight arginase activity, but the concentration increases by a factor of 10 when animals are starved for up to 4 weeks. This change in enzyme activity can be correlated with the previously described alteration in the importance of urea as a metabolic product during inanition.

Cohen and Lewis (1950) also utilized an injection technique to study the method of formation of urea within the earthworm. They inserted a fine hypodermic syringe into the gizzard of anaesthetized animals and injected amine compounds such as ornithine, arginine, citrulline, glycine, glutamine, alanine, histidine and hydantoin into the gut. The excretion of urea is affected only by arginine and citrulline when used alone. But if citrulline is injected simultaneously with another source of nitrogen such as alanine or glutamic acid, urea is formed in even greater amounts.

Therefore, although Heidermanns found no evidence for the presumption that the ornithine cycle is active in earthworms it now appears that such a cycle does in fact exist in these animals. Citrulline and arginine are both intermediates in this system, and arginase is the enzyme necessary for the breakdown of arginine to form urea and ornithine. These three steps now seem fairly well established, and further evidence on this point will be presented later. The observation that the major part of the arginase activity is located in the intestinal wall rather than the body wall is suggestive of the role of chloragogen cells in excretion. Heidermanns (1937) termed the chloragogen tissue "the central organ of urea metabolism". Needham (1960) has provided confirmation of the presence of arginase in *Lumbricus terrestris* and *E. foetida* tissues, and in the greater activity of gut preparations in this respect compared with body wall preparations. The activity of the enzyme varies in direct relationship to the urea excreted under feeding and fasting conditions.

## The Chloragogen Tissue

The function of the chloragogen tissue in the metabolic economy of the earthworm has been a vexed question for many years. Opinion has wavered between two extremes. First, it has been argued that the position of these cells, closely applied to the coelomic surface of the intestinal wall, is ideal for the development of a liver-like function. Food substances absorbed through the gut wall can be transferred into the chloragocytes, which by migration may transport the nutrients through the body or which may release these substances into the blood stream or coelomic fluid, controlling the amount in circulation by homeostatic means.

The second theory holds that these cells are excretory in function gathering waste products, which are presumably circulating freely in the blood, or coelomic fluid, concentrating them, transforming them when necessary into urea, or ammonia and then the cells detach from the gut wall to autolyse in the coelom from whence the fluid is voided via the nephridia. As high levels of urea and ammonia have been reported for both blood and coelomic fluid it is

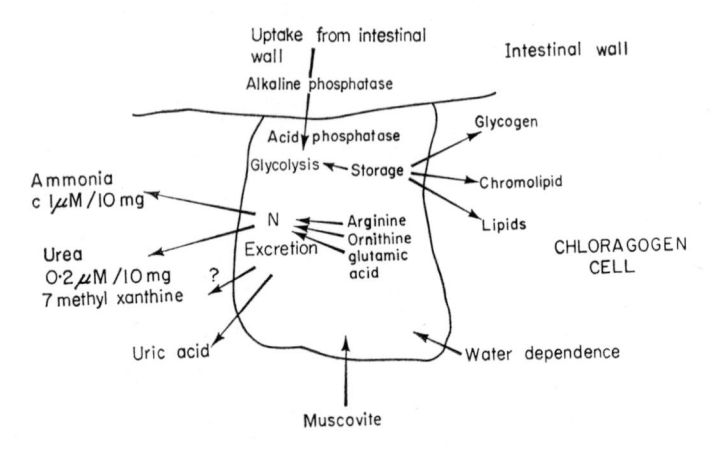

FIG. 18. Role of chloragogen cell in metabolic relations of the earthworm.

possible that both fluid systems serve to bring waste products to the nephridia, one from the tissues by means of blood vessels, and the other from the internal coelom. In earthworms with nephridia that are closed internally, of course, it is not possible for the coelomic fluid to play a part in excretion by simple filtration, only by bathing the nephridia so that diffusion can occur across the nephridial wall into the lumen.

What actually happens in the chloragocytes is probably a mixture of both the above points of view as indicated by the work of van Gansen (1956, 1957b, 1958) and Roots (1957, 1960) (Fig. 18) with which we shall be mainly concerned here, although Liebmann

(1942 a, b) has much interesting work on the nutritive functions of chloragogen which is dealt with in the section on regeneration.

The presence of food substances such as glycogen and fat has been reported at intervals since early this century, but the most exhaustive study made so far is that of van Gansen (1956). The most important food reserve, if bulk be the criterion, is glycogen. In fresh feeding *A. caliginosa* a concentration of 62·3 $\mu$g/mg dry wt. of intestine is obtained, a quantity which decreases drastically to 33·6 $\mu$g/mg dry wt. after the animals have been starved for one month. Lipid substances, also possible energy stores, are found in two forms, lipids and fats, the former remaining unaffected by starvation, the latter being depleted. Alkaline phosphatase, an enzyme often found in rapidly metabolizing tissue, is not present, its place being taken by an acid phosphatase. The probability that the materials within the chloragocytes originate in the intestinal wall is suggested by the occurrence of large concentrations of alkaline phosphatase in the distal ends of the intestinal cells lying alongside the chloragocytes, indicating the active transport of absorbed substances across the cell boundaries of this region.

The chloragocytes contain many obvious granules. Those at the periphery of the cells stain intensively with Sudan black and histochemical methods (PAS, PFAS tests) show that phospholipids are present, and contain ethylene groups. The typical brown-yellow colour of the chloragocytes is imparted by a chromolipid, the alimentary origin of which is not immediately apparent, but which is possibly an oxidized phospholipid which polymerizes with purines within the cell (van Gansen, 1957b). Roots (1960) has discussed this chromolipid and shown that it consists of a phospholipid allied with a yellow-brown spectrally uncharacterized material which she thinks may indicate that the pigment is a very complex mixture of substances in equilibrium. Other inclusions within chloragocytes include mineral aggregates, identified in electron microscope and X-ray diffraction studies as muscovite or mica (van Gansen and van der Meersche, 1958). Roots (1960) finds that not all the inorganic ash of the cells is due to silica and believes that where this substance is present it is due to accidental penetration from the lumen of the gut. The siliceous granules are not to be found universally in the chloragogen.

The facts mentioned above are all in support of the theory that chloragogen tissue is active in the metabolism of the animal, absorbing, transforming and mobilizing energy sources. But although the tissue is ideally placed for the absorption of material from the gut wall it is not easy to see how it is made available for the other tissues of the body. Three possible methods of distributing the contents of these cells are as follows: first, it is thought that chloragogen cells are able to leave the site of origin on the intestinal wall and to wander freely in the coelomic cavity. They are able to pass from segment to segment and can be accumulated at situations wherever necessary. For example it has been noted (Liebmann,

TABLE 9

NITROGENOUS COMPOUNDS PRESENT IN URINE, BLOOD AND COELOM OF THREE EARTHWORM SPECIES ($\mu$g/100 ml)

|  | P. posthuma | | | Lumbricus and Allolobophora | | |
|---|---|---|---|---|---|---|
| $NH_3$ | 2·7 | 4·0 | 2·7 | 3·22 | 1·56 | 2·04 |
| Urea | 3·2 | 2·6 | 2·5 | 7·46 | 5·47 | 5·91 |
| Creatinine | 0·5 | 3·5 | 2·7 | 1·6 | 5·4 | 6·6 |
| Protein | 30·0 | 3640·0 | 480·0 | | | |
|  | Urine blood coelom Bahl, 1947b | | | Urine blood coelom Abdel-Fattah, 1957 | | |

1942 a, b, 1946) that when a missing section of the body of an earthworm is regenerated the cicatrice area becomes packed with a mass of chloragogen tissue within a very few hours after wounding, and that these cells break down in the wound area and release their contents into the coelom. Presumably the actively proliferating cells repairing the scar and regenerating the new tissue are able to utilize the materials thus brought into the region.

Secondly, in unwounded animals the chloragogen cells also wander in the coelom, and as they do so they autolyse and the inclusions, glycogen, fats, etc., disappear into solution in the fluid. The bodily remnants are phagocytozed by amoebocytes. Quite what happens to the chemical substances thus released is as yet untraced. Some facts are known about the concentrations of substances within the coelom (Table 9) but in general the higher molecular weight organic compounds such as glucose, proteins and amino-acids are at a very low concentration compared with the blood values of the same materials. On the other hand there is less discrepancy between the quantities of nitrogenous substances in blood and coelomic fluid. This may mean that the nutrient materials are speedily removed from solution by the tissues bathed by the coelomic fluid, and consequently the concentration of these substances always tends to be low. Alternatively the chemicals may be re-absorbed into the blood system which then carries the necessary raw materials to the tissues. The blood system then, is our third possibility as a transport mechanism. Since the intestinal wall is very well provided with a plexal system of blood vessels it is easy, but unproven, to visualize the likelihood of food substances being absorbed from the gut into the blood system and distributed by this system as it wends its way round the body. Perhaps the chloragogen is indeed liver-like, acting as a homeostatic device to maintain a constant level of circulating substances. It has the added advantage of being a mobile liver and able to migrate to areas of great need.

The excretory function of the chloragogen cells for a long while rested upon a report by Willem and Minne (1900) who considered that the granules contained within the cells are composed of guanine, an end product of nitrogenous metabolism. This has been brought into doubt by Peschen (1939), Abdel-Fattah (1955), van Gansen (1956) and Roots (1957). Roots (1957) doubted the importance of chloragogen tissue as a site for nitrogenous metabolism. She isolated the granules of the cells, and analysed them to find a composition of C, 43%; $H_2$, 6%; N, 4%; P, 3·5% and S, 1%; as guanine contains 60% N it cannot be present, and it seems unlikely that other nitrogen excretory compounds are to be found as granular inclusions in the cell. But although the more complex granules may not be present there is adequate evidence that other simpler nitrogen compounds are present, and these would have been overlooked in Roots' (1957) work, since she was not concerned with materials in the liquid phase. A report that guanine is also found in the walls of the nephridia has been discounted by Bahl (1947b).

The discrediting of guanine as an excretory product has been followed by the demonstration of another purine base in these cells. Van Gansen (1956, 1957) working on the chloragogen of *A. caliginosa* found that a purine derivative, having an absorption spectrum different from that of guanine, but similar to that of heteroxanthine, a product commonly found in mammalian urine, is present in the chloragogen (Fig. 19). Heteroxanthine, or 7-methyl-xanthine, may thus prove to be of importance in the removal of one class of chemical substance from the body. This suggestion has been criticized, however, by Abdel-Fattah (1955), and also by Roots (1960) who found no trace of purine in the chloragogen of *L. terrestris*.

Though these observations are interesting they cannot be the major concern when considering the excretory role of chloragogen since these are at best minor fractions of the urine produced, and may account for the so-called creatinine fraction of Bahl (1947b).

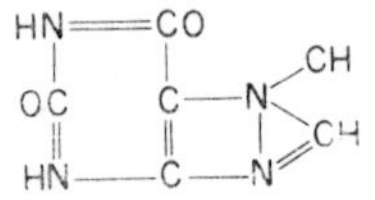

Fig. 19. Formula of heteroxanthine.

But urea and ammonia, as we have seen, are the most important excretory products, and these must be represented in the chloragogen if it is the site of origin of these substances.

In fresh chloragocytes obtained from *A. caliginosa* ammonia is found in large quantities, amounting to 0·60–1·36 $\mu$M/10 mg dry wt., whereas the urea content is negligible. After starvation the ammonia content is unaffected but urea now appears within the chloragogen at a level of 0·07–0·35 $\mu$M/10 mg dry wt. At the same time the nephridia hold 0·12–0·57 $\mu$M/10 mg dry wt., and the intestinal wall 0·12–0·18 $\mu$M/10 mg dry wt. (van Gansen, 1958). The transition to urea as a major excretory product after starvation has already been noted but the complete absence of urea from chloragocytes in fresh animals is not easily explained, although the amounts within the cells must be very small.

As mentioned previously, doubt has been cast upon the functioning of the Krebs–Henseleit cycle by Heidermanns (1937) as a result of his work on the effects of amines on the nitrogenous metabolism of tissue homogenates. No information was available to him regarding the actual amines represented in the cells, but recently paper chromatographic studies have been made of the chloragogen cells. Cells isolated from fresh animals possess three amino-acids, ornithine, glutamic acid and occasionally arginine. Glycocol and kynurenine, a breakdown product of tryptophan in mammals, are also found. A period of starvation leads to the disappearance of ornithine, glutamic acid remaining and arginine being

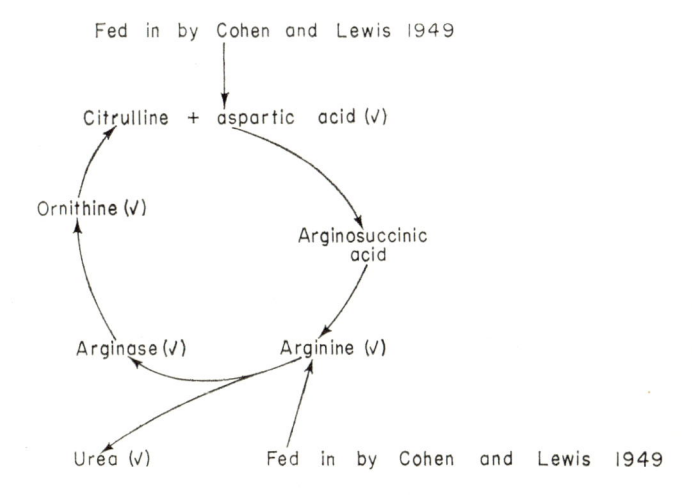

FIG. 20. Ornithine Cycle indicating components demonstrated as being present in or having an effect when fed to earthworms.

found now in all the animals studied instead of one or two. Citrulline is not present at either stage. The Krebs–Henseliet cycle of reactions (Fig. 20) involves the interaction of ornithine, arginine and citrulline. Glutamic acid is rapidly interchangeable with aspartic acid in mammals, but this reaction has not been shown in annelids. It also, by reason of combining with ammonia to form glutamine, can act as a carrier for this toxic waste product.

It can thus be seen that part of the cycle is present, but the absence of citrulline is confusing, especially so since Cohen and Lewis (1949b) report that the urea production of the earthworm rises when this amine is injected into the alimentary canal, and rises further still when another nitrogenous compound is injected simultaneously. Arginine and alanine injected into the alimentary canal can both increase the excretion of urea (Abdel-Fattah, 1955). The only enzyme yet demonstrated in this cycle is arginase (Cohen and Lewis, 1950), but it is possible that citrulline is degraded very rapidly and thus is always at a minimal level in the cells, accounting for its non-demonstration. On the other hand some completely unexpected detoxication mechanism may be in force, and Needham (1960) suggests that the cycle may be halted at the ornithine stage, thus accounting for the lack of citrulline. The actual role of the chloragogen cell in the nitrogenous metabolism of the earthworm has recently been questioned in work done by Needham (1962). The part played by the enzyme arginase in the Krebs–Henseleit system has been mentioned above, and it has been shown to occur in extracts of chloragogen cells. The results of Needham (1962) indicate that the occurrence of arginase activity along the body of the earthworm is the mirror image of the presence of chloragogen, and it seems possible that this enzyme is associated with the gut wall.

From the observation discussed above it can be reasoned that the chloragogen cell is indeed liver-like (Roots, 1960). It is evidently involved in storage and metabolism as shown by its vast glycogen reserve and fat store, the concentration of acid ATP'ase indicates that it is a highly active tissue, and the concentration and variation of ammonia and urea content shows that it is taking part in detoxication mechanisms. It would be interesting to know the effects of removal of the chloragogen tissue on an oligochaete.

*Removal of waste Products*

If then, as it seems likely from the analyses above, the ammonia, urea and other nitrogenous compounds of the urine are formed in the chloragogen cells, how are these substances transported to the organs of excretion, the nephridia that open through the body wall or into the gut in some species?

Estimations made of the nitrogenous substances of the coelomic fluid and of the blood show that in both these systems there is less ammonia and urea than is available in the excreted urine, but that creatinine (? other guanidine) is at a lower level in the urine.

Protein is at a very much reduced level in the urine, and a great deal of what is present probably arises as the mucus secretion of the body wall (Table 9). It is, therefore, obvious that a filtering process is going on somewhere in the body.

Waste materials can arrive at the nephridia by one or more of three methods: via the blood system to the walls of the excretory organs, by the chloragogen cells that disintegrate or via the coelomic fluid. The disengagement of chloragogen cells from the surface of the alimentary canal has been described by Cuénot (1898) and by Liebmann (1946); the cells fall into the coelom and there disintegrate to release the contents into the fluid of the coelom, but Abdel-Fattah (1955) does not think that such actions occur. Earthworms allowed to feed upon a mixture containing iron saccharate absorb it into the alimentary canal walls, and it is also traceable in the chloragogen cells, but in no case free or in coelomic corpuscles. It is thus unproven that the chloragogen does indeed break down in the coelom, and in so doing releases excretory products for removal via the nephridia.

On the other hand, urea and ammonia are formed in the chloragogen, and find their way somehow into the coelom. This may occur by diffusion from the chloragogen, or by filtration from the blood system. The latter idea would require some special mechanism for allowing the passage of only nitrogenous materials. Only the blood contains large amounts of glucose, amino-acids and fats, but both blood and coelomic fluid contain ammonia and urea in similar quantities (Bahl, 1947). Abdel-Fattah (1955) believes that the blood system is the transport system for the waste products of tissue metabolism and filtration occurs at the nephridia. This would seem to be the obvious method in oligochaetes with internally closed nephridia e.g. *Megascolex caeruleus* or *Hoplochaetella*, but in species like *Lumbricus* the possibility of the coelomic contents being removed down the lumen of the nephridia by the beating of the nephrostome cannot be completely discounted, particularly since the chemicals are in solution in the coelom and are not contained within the cellular entities.

*The Function of the Nephridia in Excretion*

The nephridia are arranged segmentally within the oligochaete body and good descriptions of the various modifications and structures are available from Bahl (1947), Goodrich (1945) and Stephenson (1930). See also the section on water relations and the work of Ramsay (1947, 1949).

Basically the nephridium is a tube, often open at both ends, internally and externally, and lined with cilia in tracts along its length. There is a good blood supply to this organ. The tube is specialized into distinct regions and the function of these regions is only partially known.

The many ciliary tracts that line the tubular nephridium are continually beating such that fluid passes along the lumen from the internal end to the external end. In those nephridia that are closed internally the tubular contents must obviously arise by filtration through the walls. In species having nephridia opening internally the contents may arise equally or unequally from the coelom and blood supply. If the majority of the fluid comes from the coelom then the chemical composition must be changed considerably during its passage along the tube. If it originates from the blood then many of the larger molecular substances such as protein, amines and organic food substances may be precluded from passing through the nephridial wall and thus never appear in the urine at all. Electrolytes pass through the walls into the lumen in both situations.

It is known that the composition of the urine changes as the fluid passes along the nephridium, Ramsay (1949) showing that the osmotic pressure changes are mainly a function of the "wide tube" (see Chapter VI). No work has been published to show whether salt resorption or water secretion occurs at this stage, although the phenomenon of granular resorption has been described by Cordier (1934). He found that small granules, less than 1000Å can be removed directly to the exterior via the nephridium. This compares with an upper size limit of 25Å passed by vertebrate kidney nephrons. Some granules in the earthworm, however, are accumulated by the ciliated middle tube wall, and the granules thus taken up are thought to be permanent fixtures in this situation until the end of life (Stephenson, 1930; Bahl, 1947b). The middle tube of the nephridium thus seems to act as a kidney of accumulation. It was suggested by Willem and Minne (1900) that the naturally occurring granules observed in this situation are guanine, in common with the granules of chloragocytes according to these authors, but this has been discounted by Bahl (1947b), who also disposed of the idea that they are uric acid. Guanine is not soluble in ammonia or acetic acid or dilute alkali, but the granules of the nephridium are, and they also dissolve to give a coloured solution in pyridine. This solution has absorption bands at 558 m$\mu$ ($\alpha$) and 527 m$\mu$ ($\beta$), and Bahl suggests that this is a haemochromogen deriving as a breakdown product from the blood. Further work may clarify this point since protoporphyrin (Dhéré, 1932; Laverack, 1960a) is known to be laid down in the body wall during life, giving characteristic iridescence to the integument, and this substance is soluble in pyridine. The absorption peaks differ from those of the nephridial granules. Roots (1960) has recently investigated the brownish-yellow pigment of the chloragogen cells and found that this may be a complex arrangement of substances which are soluble in ammonia and pyridine, but possess no distinct absorption spectrum. This may bear some resemblances to the nephridial material, but the question needs further work.

From his comparative studies on the blood, coelom and urine of *P. posthuma*, Bahl (1947b) concludes that nephridia carry out three functions in earthworm excretion, namely filtration, re-absorption and chemical transformation. Filtration, in the sense of pouring a liquid through a coarse sieve, but not ultrafiltration under pressure through a membrane as in the vertebrate kidney, occurs at the nephrostome. Thus in nephridia open to the coelom, protein passes into the lumen of the nephridium, and it is unlikely that it can do so from the blood vessels by ultrafiltration since the pressure within these vessels is low (*ca.* 4·4 mm Hg at rest and >9·3 mm Hg when active, Prosser *et al.*, 1950). As the protein appearing in the urine is far less than that in the blood and coelomic fluid, and much of what is present is accounted for by the secretions of the body wall, it follows that much active re-absorption of protein occurs through the nephridial wall against a concentration gradient. Electrolytes such as chloride, sodium, potassium and phosphate all diminish in quantity as they pass along the tubules since the final concentration is lower in the urine than in the coelomic fluid and blood. Although the evidence is not conclusive, Ramsay's observation that urine becomes hypotonic in the wide tube, and possibly more so in the middle tube is also suggestive of the re-absorption of substances in these regions. Martin (1957) recently discussed these results in the light of more modern knowledge in other invertebrates. He considers that if blood is filtered across the closed nephridium of *Pheretima* a large osmotic inflow of water would occur, thus forming a hypotonic urine, and causing the animal to handle up to 50% of the body volume of water in 24 hours. Bahl (1947) found a urine production of very nearly this figure (45%) in animals maintained in a water-saturated atmosphere, equivalent to an excretion of 0·3–0·4 ml/hr/worm. Chapman (1958) considers that a urine production rate very similar to this is shown by *Lumbricus* as well. If this is so then the hypotonic production may be due to flooding of the nephridium. Electrolytes may be actively re-absorbed through the nephridial wall, passing into the coelom (or blood?), and after the elapse of some further period isotonicity would again be established between coelom and blood, (Fig. 21).

Alternatively Bahl (1947b) indicates that filtration from the blood capillaries surrounding the nephridium may occur, even in situations where they are open to the coelom. This may be facilitated by the beating of the cilia within the tubules. A slight pressure would be set up in the tubule and this may act in drawing liquid through the walls of the nephridium. Carter (1940) previously considered this to be unlikely as the colloid osmotic pressure of the fluid bathing the cells is too great, but Pantin (1947) has shown that the flame cell activity of a nemertine *Geonemertes dendyi* increases as fluid passes into the animal and osmotic regulatory mechanisms come into play. This sort of action may occur in annelids as well. Indeed Roots (1955) reports that after bathing with hypotonic fluids the cilia of the nephrostome (*Lumbricus terrestris* and *Allolobophora chlorotica*) undergo a period of increased activity in rate of beat and sometimes amplitude as well. This is followed by a return to normal. In cases of hypertonicity the ciliary beat slows.

The question of re-absorption of materials from the nephridium Martin (1957) considered to be unconfirmed, particularly the removal of protein from the fluid in the lumen. He considered that the differences in blood and urine levels of protein, which are in the ratio 16:1 requires that the coelom should provide more than one-sixteenth of the urine filtered. Another objection is that, in

# NITROGENOUS EXCRETION

*Pheretima* at least, the nephridia drain, not to the outside, but into the gut and the absorption of protein may occur here rather than

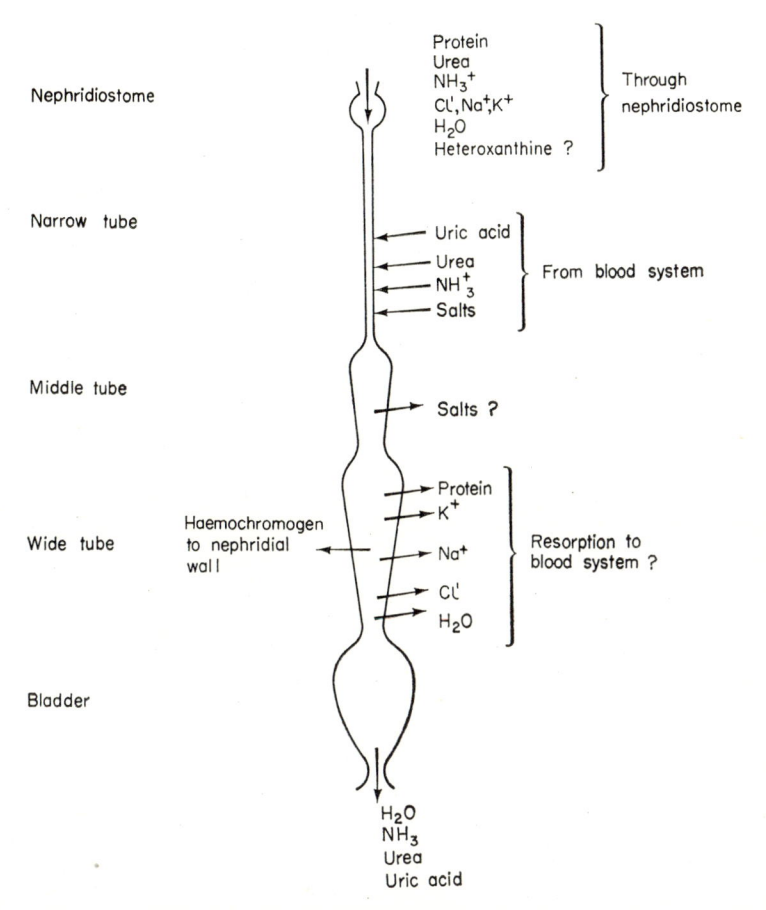

```
Nephridiostome          Protein
                        Urea
                        NH$_3^+$           Through
                        Cl', Na$^+$,K$^+$  nephridiostome
                        H$_2$O
                        Heteroxanthine ?

Narrow tube    ← Uric acid
               ← Urea            From blood system
               ← NH$_3^+$
               ← Salts

Middle tube    → Salts ?

               → Protein
                            Haemochromogen
Wide tube      → K$^+$      to nephridial    Resorption to
                            wall             blood system ?
               → Na$^+$
               → Cl'
               → H$_2$O

Bladder

               H$_2$O
               NH$_3$
               Urea
               Uric acid
```

Fig. 21. Diagram to summarize possible mode of functioning of oligochaete nephridia (*Lumbricus*).

in the nephridium. Ramsay (1949b) found that whilst a considerable coagulation occurs upon heating blood from *Lumbricus* only a very slight coagulum occurs with coleomic fluid or urine, and as urine probably derives from both sources in this species the protein content of the final fluid can be expected to be less than that of the coelomic fluid.

*Summary*

Nitrogen is lost in the form of mucus secreted by the epidermis and in urine.

Earthworms excrete ammonia, urea and also, possibly, uric acid, allantoin and doubtfully creatinine. The concentrations of these substances varies according to the diet. In animals under normal conditions of food and water intake, ammonia is the dominant constituent in urine with smaller quantities of urea; under conditions of starvation the position is reversed, urea increasing above the previous level and above that of ammonia. Some evidence is available that the chloragogen cell is involved in excretion probably via the Krebs–Henseleit cycle of reactions. The enzyme arginase, and the amino-acids ornithine, glutamic acid and arginine have been shown to be present but not citrulline. Heteroxanthine may be present in chloragogen. There are indications that re-absorption of material occurs in the nephridia but filtration from the blood and coelom may be efficient enough to account for differences in analyses of urine, blood and coelomic fluid.

CHAPTER VI

# WATER RELATIONS

THE PROVISION and maintenance of a sufficient quantity of water within the body, either as lymph, blood, tissue water, mucus, etc., is a problem that besets all animals. Each species has its own particular problems with regard to dehydration, or flooding by excess water, and each habitat has its own set of difficulties. Three broad divisions into three habitat types may be mentioned, those of the sea, fresh water and the land. Within each type there is considerable variation, of course, as for example in the gradual transition from fresh water to salt water found in the estuarine reaches of rivers, and in the passage from swampy marshland to dry arid desert found on the land surface.

Marine animals, living in a medium of high osmotic pressure, are always tending to lose water in response to the osmotic gradient that exists between their bodies and the external medium. Many marine animals remove their excretory products as simple, highly soluble nitrogenous substances such as trimethylamine oxide or ammonia, which are released into the sea. Salt-secreting cells are found in the gills of marine fish which aid in keeping the internal environment constant by removing excess mineral ions taken in through the mouth in the sea water necessary to avoid desiccation.

Fresh-water animals, however, are faced with a different problem. In their case the osmotic pressure of the body fluids within the animal is greater than the osmotic pressure of the medium and consequently water passes inwards through the body wall and incessantly dilutes the internal fluids. This situation is countered by the production of copious quantities of hypotonic urine by the animals which maintain a constant internal environment in this manner.

Terrestrial animals are faced with a similar problem to that of marine animals, that of dehydration, but for a different reason. In marine animals water is lost to the greater external osmotic pressure; in terrestrial animals water is lost by evaporation from the body surface, and by leakage from the various natural pores of the body. Conservation of water is, therefore, of prime importance, particularly in those animals living in regions where water is scarce. In many animals the excretory products have changed from the highly soluble, but highly toxic ammonia and trimethylamine oxide, in favour of the less soluble but also less toxic substances, urea and uric acid. These require less water for their removal and thus excretory processes do not denude the animal of very necessary water supplies.

Oligochaetes are found in fresh water and upon land. We thus expect to find appropriate modifications in their physiology to deal with the special strains in water balance imposed by these habitats.

As mentioned previously (p. 2) the earthworm body contains a very large amount of water. The average water content amounts to about 85% of the fresh weight of earthworms. A considerable proportion of this total is present as coelomic fluid or as blood, and a further large fraction is found within the gut in the form of secretions. Water can escape from the body via the mouth and anus, the dorsal pores and nephridia and as mucus, either singly or in combination. Very little is known at present about the water relations of fresh-water oligochaetes such as *Tubifex*, though it is to be expected that water passes in through the body wall and that the animal excretes considerable quantities of hypotonic urine in common with many other fresh-water animals, and similarly to earthworms.

The terrestrial oligochaetes *Lumbricus* and *Eisenia* have only a thin cuticle overlying a mucus secreting membrane, the epidermis. It is unlikely that the cuticle plus epidermal layer provides a serious barrier to the influx and efflux of water through the body wall, so that changes in environmental conditions can lead to considerable changes in the water balance of the animal.

Earthworms dug from the soil are not fully hydrated and will increase in weight if immersed in tap water. They may gain up to 15% in weight in 5 hours, losing it again when replaced in soil (Wolf, 1938, 1940). The loss of weight in soil is a faster process than the gaining of weight in water (Adolph, 1943). These minor weight and accompanying volume changes serve to show that under normal field conditions the earthworm maintains a water equilibrium with fair efficiency. The locomotory and burrowing efficiency of annelids depends to a large extent upon the hydrostatic skeleton of the coelomic fluid and loss of water in particular may interfere with normal behaviour. Water losses and gains of the magnitude mentioned above do not affect the normal locomotion and activity of the animal, a loss of some 18% being tolerated before overt changes in behaviour patterns are observed (Wolf, 1940). Slight diurnal alterations in body weight amounting to an average of 2–3% of the basal weight occur which have little significance, but if earthworms are kept in distilled water for up to 16 days there is, after an initial gain, a gradual loss of weight which may be due to leaching of salts from the internal fluids (Wolf, 1940).

Roots (1956) found that many species of earthworm are able to live in waterlogged soil which has standing water above it. Survival of *L. terrestris*, *A. chlorotica* and *A. longa* among others was for 31–50 weeks in such conditions, and the animals remain alive for 72–137 days without food. Many individuals in such conditions remain partly stretched above the surface of the soil in the water layer. Burrows made in waterlogged soil are not irrigated as are those of burrowing polychaetes. Roots suggests that the long survival of earthworms under water depends not upon the ability to control the water circulation of the body, but upon the ability to withstand prolonged starvation. Earthworms will leave waterlogged soil in preference for a drier site if possible.

The respiration of earthworms takes place through the body surface, without the aid of special anatomical structures. It is essential for gaseous exchange to take place that a thin film of moisture be maintained at the interface between the air and the skin, in order that oxygen may go into solution before passing across the boundary of the skin and into the blood system. Obviously if the animals live in a moisture-saturated atmosphere it is a comparatively simple task to maintain a layer of water over the skin since evaporation will be at a minimum. Soil conditions are rarely stable with regard to moisture content, however, and if they become dry then earthworms lose moisture from the body. An attempt is made to keep the surface moist and if water is continually lost the animal loses weight. Up to about 75% of the total body water may be lost without causing death (Roots, 1956, Schmidt, 1927). As dehydration becomes more pronounced a series of behavioural changes occurs; at first there is a prostomial reaction, when it is noted that the ability of a substrate to induce dehydration of the body calls forth a "dehydration tropism". Parker and Parshley (1911) found that the response was localized on the prostomium, the removal of which stopped the reaction, the response being shown again when the prostomium regenerated. This tropism is followed by a period of rolling, and then coelomic fluid is expelled from the dorsal pores in an endeavour to moisten the body surface, but if the desiccation is unrelieved this is followed by rigor,

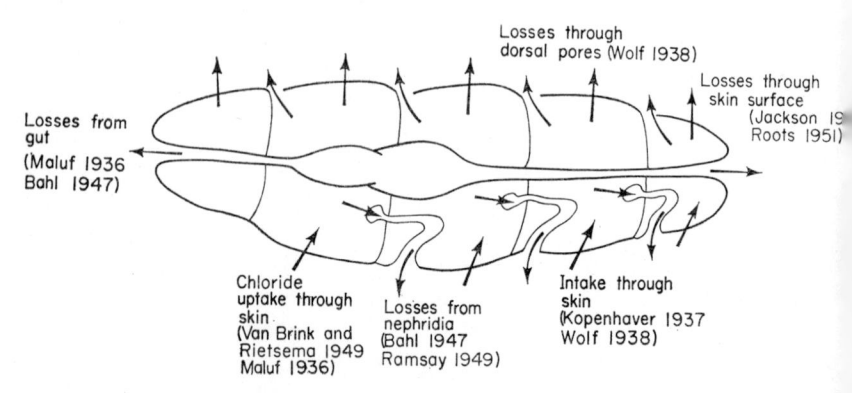

FIG. 22. Diagram summarizing routes of water loss or gain in the earthworm.

anabiosis and finally death (Wolf, 1938). Providing that irreversible internal changes have not occurred animals can be revived by submersion after very large water losses.

Water is lost readily through the skin, and it can be gained by the same pathway from the external environment with equal facility. *L. terrestris* ligatured at both anterior and posterior ends prior to immersion in water gains weight, about 7%, and then maintains a constant weight level (Maloeuf, 1940 a, b). Not only can water be taken in via the body wall, so also can salts. Chloride ions disappear from dilute salt solutions containing worms due, it is thought, to active uptake of the ions (Maloeuf, 1940a; Van Brink and Rietsema, 1949).

If earthworms are allowed a choice between water-saturated or moist, air-filled soil, the very wet areas are avoided by *Dendrobaena subrubicunda* and *L. terrestris* but *A. longa*, *A. caliginosa* and *A. chlorotica* are sometimes found in the waterlogged soils. This last-named species is interesting as it exists in two colour forms, one pink and apparently living in drier soils, whilst the other is green and often found in very wet areas, such as on the muddy floor of Lake Windermere (Kalmus *et al.*, 1955). In laboratory studies no difference was found between the soil preference of these two forms (Roots, 1956).

The avoidance reaction of earthworms to water has been suggested as the causal agent in the well-known evacuation of burrows by certain species after heavy rain (Roots, 1956) first noted by Darwin (1881), but recordings of the electrical activity in the segmental nerves of *L. terrestris* and *A. longa* indicate that water does not stimulate sense organs in the body wall (Laverack, 1960b), although this does not preclude the possibility that water entering the body cavity through the skin dilutes the coelomic fluid bathing the central nerve cord, increasing the activity within the nerves in a manner analogous to that noted in slugs (Kerkut and Taylor, 1956). In this connection it has been noted that anaesthesia or extirpation of the nerve cord leads to a breakdown of the osmoregulatory powers of another annelid, the leech *Hirudo medicinalis*, and that this appears to be a function of the whole nerve cord and not a particular portion (Roşça, Wittenberger and Rusdea, 1958), but preliminary results on *L. terrestris* have been negative (Laverack, 1960 unpublished).

The ways in which an earthworm maintains a constant, or reasonably constant, internal medium has been the subject of some disagreement until recently. The actual value to be attached to the osmotic pressure of the coelomic fluid has been reported to vary widely, and this may be a reflection of the bodily conditions of animals taken fresh from soil sites of different properties. The freezing-point depression of the coelomic fluid in *Lumbricus* is approximately 0·31 °C (Adolph, 1927 and Ramsay, 1949a), whilst that of *Pheretima posthuma* lies between 0·28 and 0·31 °C (Bahl, 1947a). The osmotic pressure of the blood of *Lumbricus* is slightly below that of the coelomic fluid (Ramsay, 1949a) but in *P. posthuma* the freezing point depression of blood is 0·4–0·5 °C indicating that blood has a greater chemical content than the coelomic fluid. Analysis of the chloride content of the body fluids reveals that only about half of the osmotic pressure in *L. terrestris* can be accounted for as chloride, the remainder possibly being made up by organic substances (Ramsay, 1949a). The chloride content of the blood of *P. posthuma*, however, is only half that of *Lumbricus*, so it would seem that in this species organic substances such as glucose, protein and other large molecules must provide an even greater proportion of the total osmotic pressure.

The earliest work on the maintenance of constant internal conditions is due to Adolph (1927) who believed that all the water exchange of earthworms is carried out by two agencies. Water enters through the skin, a process which goes on very easily, and is removed via the intestine. The nephridia play no part at all in excretion of water. This view was contested by Wolf (1940) who found that the loss in weight sustained by worms when handled was due mainly to expulsion of fluid from the nephridiopores, and Adolph (1943) later accepted this observation.

A possibility that both types of excretory process may occur was suggested by Maloeuf (1939). He considered that water entered the body across the body wall, was stored in the gut, and later removed via both the nephridia and the gut. The gut was thought to be particularly important in conditions which place the nephridia under great stress, such as when the animal is placed in water and is subjected to continual flooding by water entering by diffusion. It was shown that if the gut was ligatured so that the contents could not be removed the animal swelled as more and more water diluted the coelomic fluid. Some control was exerted by the nephridium and Maloeuf (1939) thought that the urine must be hypotonic to the body fluids in order to remove the vast excess of water, but he was not able to check his ideas by actual measurements. Later Maloeuf (1940b) obtained figures for the osmotic pressure of the gut contents and found it to be only slightly less than that of the body fluids and thus was probably not after all involved in water excretion; the slight excess of water being used as lubrication for the soil particles within the gut.

It must be remembered that the environment of the earthworm is terrestrial. Fresh-water oligochaetes, about the excretion of which nothing is known, might be expected to be always pumping water from the body as fast as it diffuses inward across the body wall, with the production of a copious hypotonic urine. But an earthworm finds itself between two stools, being neither completely immersed in water nor living in rigidly arid surroundings involving it in strict water conservation. The environment can vary anywhere between these two extremes.

On average, however, it must be considered that the normal environment of the earthworm is midway between that of a fresh-water animal and that of an air dweller. Under such conditions Bahl (1945) considered that in *Pheretima* the nephridia function adequately as volume and osmoregulatory organs since the volume of water involved is not excessively large. But when a terrestrial earthworm is kept in water like a fresh-water animal, the large amounts of water passing across the skin cannot be eliminated by the nephridia and the gut is then brought into action, removing excess water through the anus and the mouth.

Stephenson (1945) obtained evidence that considerable salt regulation is carried out during the removal of water. He placed *Lumbricus* in dilute salt solutions, finding that the internal chloride concentration remained above the external concentration. In other experiments he put earthworms into concentrated salt solutions and noted that they maintained a low internal concentration relative to the external medium. On the basis of this he concluded that the earthworm is not like a truly fresh-water animal "as far as its osmotic relations with the environment are concerned".

This work was repeated by Ramsay (1947, 1949) in a series of very elegant and careful experiments. Ramsay utilized fine micropipettes which he inserted into the nephridiopores of *Lumbricus* to collect urine. He was then able to compare the osmotic pressure (O.P.) of this fluid with that of the osmotic pressure of the coelom and of the blood. He also confirmed that chloride accounts for only 50% of the total blood O.P., and that blood is just hypotonic to coelomic fluid ($\Delta = 0.31$ °C for coelom). He kept his animals in media having salinities between 0·025 and 1·27% NaCl, and found that as the concentration of the medium increases so the osmotic pressure of the body fluids also increases, remaining always greater than that of the medium. The chloride content increases proportionately but is less than that of the medium when the latter exceeds 0·35% NaCl. The urine obtained from the nephridiopores is always hypotonic to the body fluids except in the most concentrated media ($>1.0\%$ NaCl) (Table 10). It is thus clear that the earthworm has no powers of osmoregulation in concentrated media, although the chloride data alone may suggest such a course. In fact survival in very concentrated media is often not for very long, and urine production is very limited.

It is evident then that the earthworm, so far as is known at present, functions like a fresh-water animal, albeit with some anomalies. The maintenance of an internal O.P. above that of the environment, save in concentrated media, the elimination of hypotonic urine, the absorption of salts from very dilute solutions,

TABLE 10

TOTAL OSMOTIC PRESSURE AND PROPORTION DUE TO CHLORIDE AS % NaCl IN RELATION TO THE CONCENTRATION OF THE EXTERNAL MEDIUM (From Stephenson, 1945)

| Medium | Urine | | Coelom | |
| % NaCl | O.P. | Cl | O.P. | Cl |
|---|---|---|---|---|
| 0·025 | 0·10 | 0·02 | 0·53 | 0·27 |
| 0·65  | 0·23 | 0·10 | 0·85 | 0·43 |
| 0·75  | 0·30 | 0·14 | 0·95 | 0·48 |
| 1·27  | 1·37 | 0·50 | 1·40 | 6·75 |

and the volume of urine production (estimated at 60% of the body weight in 24 hours by Wolf (1940)), all point towards a fresh-water type of excretory pattern. The fact that the chloride concentration is kept below that of the external medium does not require the postulation of an active process of salt removal against a concentration gradient. In media over 0·35% NaCl salt doubtless diffuses in, possibly coupled with the active uptake process involved in dilute solutions. But water also tends to enter since the internal O.P. is higher than the external, and as this latter process probably occurs more swiftly than the former, due to the greater permeability of cells to water, the effect is that of dilution of the coelomic fluid. The animal can thus maintain the chloride lower than in the medium and at the same time eliminate a hypotonic urine.

There are a few conflicting facts affecting the simplicity of the picture of passive entrance of water under osmotic pressure, passive and active entry of salts with or against the osmotic gradient and elimination of hypotonic urine. For example it is known that when earthworms recover in water after considerable desiccation, the rate of intake of water increases many-fold at deficits that do not even double the osmotic pressure of the body fluids (Adolph, 1927). The volume of the body also holds an anomaly when compared with the O.P. of the medium for Stephenson (1945) found that the body volume is at a maximum at an external concentration of 0·35% NaCl ($\Delta$ of body fluids = 0·31%) and declines as the medium becomes more or less concentrated. This is unexpected on the simple system outlined above. Ramsay (1949a) suggests that further work will indicate where the modifications in this picture are to be found.

*The Nephridium*

By the use of still more refined techniques Ramsay (1949b) collected fluid from the various tubules of the nephridium with the object of determining the site of production of the hypotonicity of the urine. The osmotic pressure of urine samples from the ampulla, the ascending and descending sections of the narrow tube, the middle tube, the mid and distal portions of the wide tube, and the bladder of the nephridia of *Lumbricus* were estimated (Fig. 23b).

Figure 23a shows the results Ramsay obtained. The narrow tube probably contains fluid isotonic with the coelomic fluid that normally surrounds it. The diameter of this tube becomes greater almost immediately if the bathing fluid is diluted, suggesting that ingress of water is very rapid, but whether it goes through the nephridial walls or down the nephrostome at an increased clearance rate is not obvious. The middle tube also swells when the external fluid is diluted, but does so more slowly than the narrow tube. The wide tube and the ampulla do not change in diameter.

There is almost certainly an increased ciliary clearance rate in animals with diluted coelomic fluid. Worms kept in 1·4% NaCl for one week had nephridia that were small and shrunken and with the ciliary tracts almost stationary. The bladders were empty. When the bathing saline was replaced by 0·45% NaCl first the narrow tube and later the middle tube opened and the cilia started to beat again. The bladders became filled after about 30 minutes (Ramsay, 1949b).

Ramsay's results show that almost certainly no change occurs in the composition of the urine within the narrow tube. The osmotic pressure of fluid here is unchanged from that of the external fluid, indicating perhaps that water and salts pass in both directions across the wall with equal facility, or that filtration is occurring

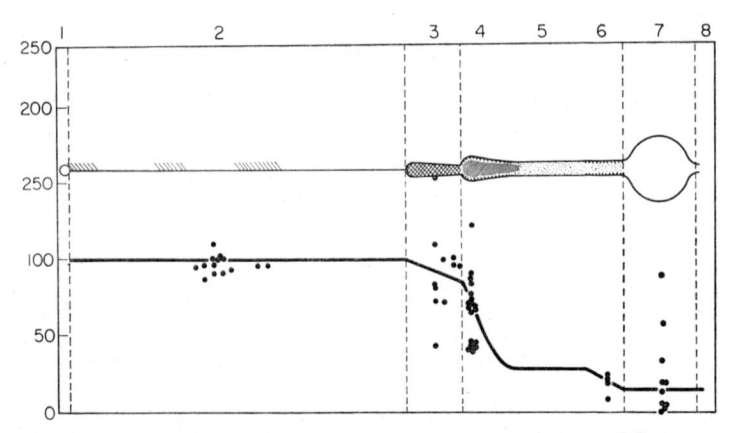

FIG. 23a. To show the osmotic pressure of urine at different levels in the nephridium. The osmotic pressure of the ringer surrounding the nephridium has been equated to 100. 1. Nephridiostome. 2. Narrow tube. 3. Middle tube. 4. Wide tube proximal. 5. Wide tube middle. 6. Wide tube distal. 7. Bladder. 8. Exterior (from Ramsay, 1949b).

across the wall of the blood vessels around the organs into the lumen of the tubule, or that coelomic fluid passes directly through the open nephrostome and from there is passed along to the middle tube. In this region more variable results were found, but in general a slight decrease in osmotic pressure was found. This drop was more pronounced, however, in the next section of the nephridium, the ampulla of the wide tube. In this region the osmotic pressure may be as much as halved compared with the external medium. At the distal end of the wide tube the osmotic pressure is even less, suggesting that the property of forming a hypotonic urine is carried out along the entire length of the wide tube. It is also possibly a function of the middle tube since there is a certain amount of variability of the results. The bladder seems to be solely a collecting and storage organ for the formed urine; no changes in

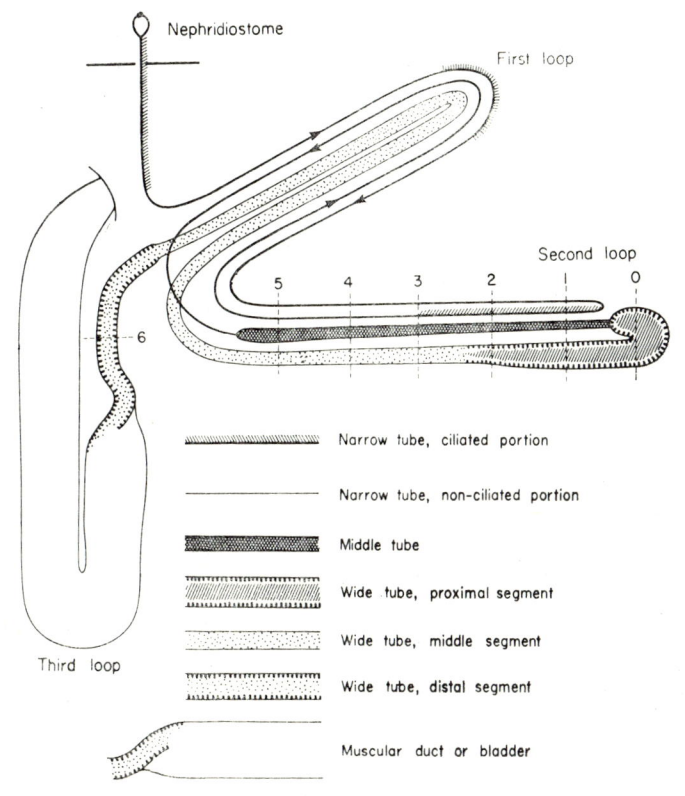

FIG. 23b. Diagram of nephridium of *Lumbricus* (from Ramsay, 1949b) showing regions sampled with micropipettes.

the concentration of the contents being found in relation to the wide tube.

There is no knowledge available at present to say whether the changes in urine concentration noted in the wide tube are due to an influx of water into the tube, to removal of salts from the lumen by active resorption into the blood system, or whether it is a combination of both of these phenomena. There is some evidence given by

Cordier (1934) that resorption of particles occurs in the nephridium of oligochaetes. Particulate matter injected into the coelom of the worm is found to accumulate in the wall of the middle tube, the finer particles being taken up at the proximal end, the larger particles at the middle, of the middle tube. Cordier postulated that the process of water resorption may occur in the wide tube of the nephridium by analogy with the distal convoluted tubule of the vertebrate kidney which has a similar cytological make-up. Ramsay (1949b) finds that none of his experiments refute this idea, but neither do they make it any more likely. Similarly he throws little light upon the likelihood of salt-resorption taking place in the same region. And there for the moment the matter rests.

*Nephridial Activity under Changing Conditions*

More information about the activity of the nephridia under changing external conditions has been presented by Roots (1955, 1956). She isolated nephridia completely from the body into saline solution, either frog Ringer diluted with an equal volume of $M/400$ $NaHCO_3$ or 66% frog Ringer alone. Though survival in these solutions was often prolonged, up to 22 hours, it is perhaps unfortunate that no one has yet undertaken the plea by Ramsay (1949b) that further knowledge of the ionic composition of the coelomic fluid of the earthworm is necessary in order that isolated tissue experiments may be more realistic. However, Roots obtained nephridia from *L. terrestris* and *A. chlorotica* and subjected them to media of various strengths. When the saline is suddenly diluted the nephridiostome cilia become more active, with an increased amplitude of beat, returning after a variable period to the original state. Long-term treatment with hypotonic solutions leads to arrest of the cilia, more quickly in *L. terrestris* than in *A. chlorotica*. Vesicles form in the cells of the nephridiostome under these conditions.

Hypertonic solutions, on the other hand, decrease the activity of nephridial cilia. The cilia again become active within 30 seconds after replacement in isotonic saline. These temporary changes in activity should be compared with those described by Ramsay and mentioned above.

The concentrations of media used by Roots varied from 30–300% frog Ringer (diluted with $M/400$ $NaHCO_3$) and bear little relation to such concentration and dilution of body fluids as the earthworms may encounter in nature save in extreme conditions. Consequently Roots (1956) attributes little importance to the resistance of the nephridia to changing osmotic conditions. None the less it has been mentioned above that the earthworm taken fresh from the field is not completely hydrated, but can become so under conditions when free water is available such as heavy rain, or lose water when soil humidity decreases so that some strain must be imposed upon the organs of excretion and volume control. For example Zicsi (1958) has shown that 60% of the body water is lost from many species of worms living in soils containing less than 10% water. This is an exceptionally low quantity of soil water, but such humidity levels can be reached during long dry spells of weather.

The hydrostatic pressure within the coelom of earthworms amounts to approximately 15 cm water, and the total number of nephridia to about 300. On this basis Chapman (1958) calculates that between 1/100 and 1/4 of a millilitre of fluid will be lost per hour if the nephridia are considered simply as open tubes connecting the coelom to the exterior and the animal maintains a constant internal pressure. The nephridia can of course, be sealed by a sphincter muscle when necessary, as in the passage of a peristaltic wave. The amount of fluid escaping in this calculation agrees closely with the volume of urine production observed by Wolf (1940) and by Ramsay (1949b). Thus the coelomic pressure could account for much of the urine appearing at the nephridiopores, but Ramsay noted that collection of urine occurs in pipettes. This involved an active excretion of fluid so evidently some other force combines with coelomic pressure to account for the formation of urine in the normal animal.

*Aestivation*

In some species of earthworm more than others, the variations in external water have pronounced effects upon the behaviour of the animals. Such worms as *A. longa* and *E. rosea* undergo a period of aestivation. This occurs during the summer months when the animal becomes isolated, curls into a small knotted ball in a pocket of air in the soil and surrounds itself with mucus secretions which dry to form a coating on the inside of the pocket. The onset of aestivation, or diapause, is associated with low soil humidities, and it is possible to prevent its appearance by keeping earthworms in a humid atmosphere. *A. longa* kept at 18 °C in high humidity with plenty of food shows cyclic clitellar development and reproduces normally. Slow dehydration and lack of food is followed after 4–6 weeks by diapause, the animals coiling up, becoming colourless and isolating themselves in a pocket of earth. For the next 2 months the animals cannot be awakened, spontaneous wakening occurring at the termination of this state. A great deal of weight is lost during this period, presumably due to loss of water by evaporation and food reserves by metabolic breakdown.

It is also possible to induce aestivation in the winter by prolonged desiccation, but although humidity conditions are of great importance in controlling diapause, other factors play a considerable part. In view of the recent research into neurosecretion associated with reproductive and regeneration capacity it is highly likely that internal cyclic factors are concerned in the regulation of aestivation (Michon, 1949, 1954).

*Summary*

Earthworms freshly dug from the soil can take up water, losing weight again when returned to the habitat. Many species can remain alive in waterlogged soil under standing water. Severe loss of water can be withstood and the prostomium appears to be the site of response for a dehydration tropism. Water passes rapidly across the body wall in both directions, and there is evidence that mineral salts may do likewise. In its normal water relations the earthworm functions as a fresh-water animal with one or two modifications. It excretes a copious hypotonic urine and this is modified in its composition during its passage through the nephridium. Ciliary activity within the nephridial tubules increases when the animal is immersed in water of low salt concentration, thus diluting the coelomic fluid. Some earthworms undergo a period of aestivation when the external humidity drops. This may be governed by a neurosecretory process.

CHAPTER VII

# RESPIRATION

THE process of respiration involves a multitude of factors. At the morphological level there is the necessity for providing a surface across which diffusion of gases can occur e.g. lungs and gills, and the necessity for providing, even in the most arid situations, a continuous layer of moisture into which the gases can dissolve is essential. In many cases it is also necessary to provide a current of water or air over the exchange surface as is accomplished by ventilation. Internally there needs to be, in the more highly organized animals, a means of transporting the dissolved gases to the appropriate sites for utilization and removal. This involves the dissolution in fluid circulating in vessels, or passage of air in trachea. The process may be assisted by the occurrence of a pigment that combines readily with oxygen to carry it to the tissue, there to release it as required. The removal of carbon dioxide may also be accomplished by special devices. Last, but not least, at the cellular level there are all the many enzyme reactions that go to provide energy for metabolism synthesis and excretion and which rely ultimately on the provision of oxygen and removal of carbon dioxide.

It is obvious that there will be many variations upon these things, dependent upon the complexity, activity and ecology of the species in question. What suffices for the amoeba, a straightforward gaseous diffusion across the cell surface and no need for a transport system, will obviously not do for a fish that swims actively throughout its life e.g. *Scomber scomber*, and must maintain a current over its gills for gaseous exchange followed by a highly organized vascular system involving a pumping mechanism, valves and a respiratory pigment.

The majority of oligochaetes fall midway between these two extremes. For while there is often no specialized gas exchange organ there is a transport system and a respiratory pigment.

84   THE PHYSIOLOGY OF EARTHWORMS

In a few cases, for example *Dero* and *Branchiodrilus*, there are gills presenting a large surface area for the purposes of gas exchange (Fig. 24). This situation contrasts greatly with the great profusion of

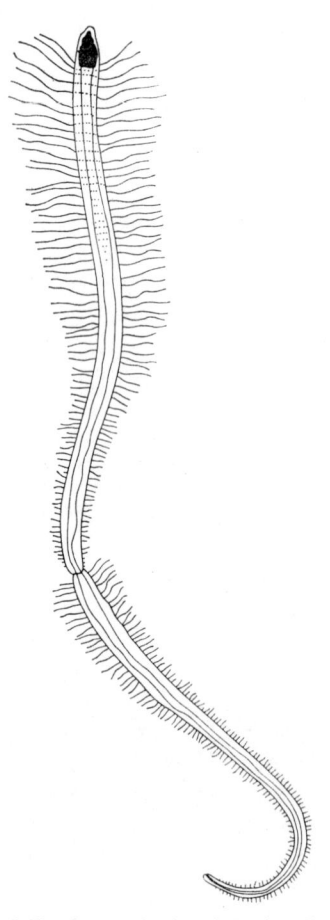

FIG. 24. *Branchiodrilus hortensis* showing the distribution of the gills (from Stephenson, 1930).

gill structures in polychaetes. In earthworms about which we have most knowledge, the only structural specialization consists of looped and branched capillary extensions of the blood vessels in the body wall (Fig. 25). These contain haemoglobin and diffusion accounts for the passage of gases to and fro.

One species of oligochaete, however, is worthy of special mention, namely *Alma emini*, a glossoscolecid oligochaete found in the papyrus swamps of East Africa. It lives in a substratum of decomposing vegetable matter that is strongly reducing and probably devoid of free oxygen. *Alma* lives immersed in the mud and very often only the rear end projects above the surface into the air. This rear extremity is modified to form a highly vascularized lung (Fig. 26). The edges of the body fold over to form a tube in which bubbles of air are frequently trapped and taken down when the animal retreats from the surface. The haemoglobin of this oligochaete has a very low unloading tension, being oxygen saturated at a partial

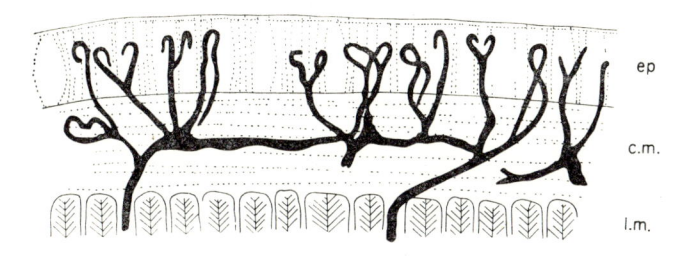

Fig. 25. Intra-epidermal capillaries of *Lumbricus*. ep = epidermis; c.m. = circular muscle; l.m. = longitudinal muscle (from Stephenson, 1930).

pressure of 2 mm Hg oxygen in the absence of carbon dioxide. In other words in the oxygen-deficient layer where the animal lives it is able to make full use of the little oxygen which is available. In an ecological habitat such as this carbon dioxide may be expected to accumulate to high levels, and it is notable that high $CO_2$ tensions are without noticeable effect on the dissociation curve of the haemoglobin of *A. emini*. Only at a tension of 200 mm Hg of $CO_2$ is the curve slightly displaced to the right, i.e. the unloading tension increases (the Bohr effect) and worms can live for up to 48 hours in water completely saturated with $CO_2$. Although laboratory studies with this species show that it can survive under completely anaerobic conditions its behaviour in mud:water culture suggests that oxygen is a prerequisite for normal existence (Beadle, 1957).

In all respiratory systems oxygen first dissolves in a watery layer covering the respiratory surface. It proceeds from there into the body by a passive diffusion, not by an active process. This has been confirmed for earthworms by analysis of the gaseous exchange in relation to body area and weight (Krüger, 1952). The surface of

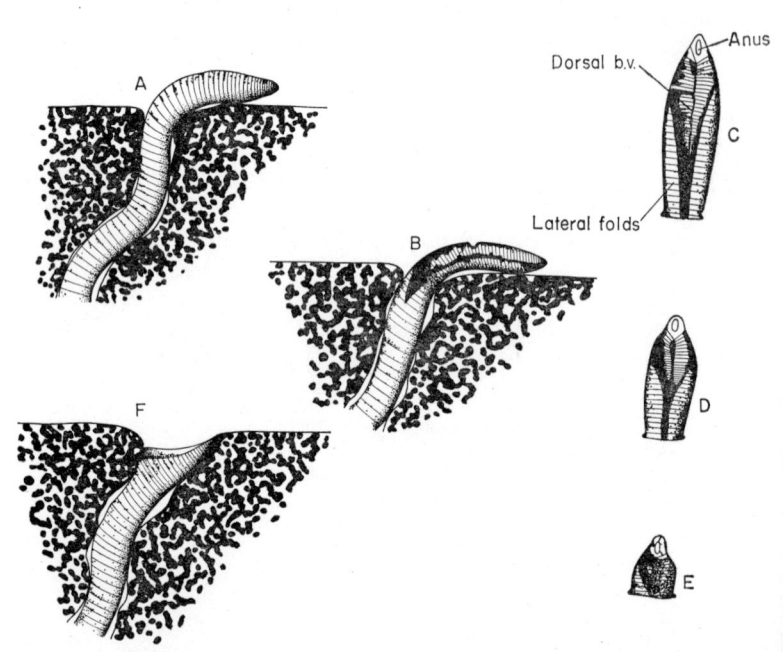

Fig. 26. Exposure to the air of the posterior dorsal surface and formation of "lung" in *Alma emini*. A. extrusion of hind end. B. hollowing of dorsal surface. C.–E. dorsal view of extruded hind end. Stages in the formation and closure of lateral fold and retreat into mud. The dorsal blood vessel and numerous lateral connections are clearly visible in this region. F. final position after retreat into mud with tubular lung open to the air (from Beadle, 1957).

the body is kept moist by the continuous secretion of mucus which also serves a purpose as a lubricant during locomotion, and cementing soil particles in the burrow. Desiccation leads to increased mucus secretion, and in the last event to an expulsion of fluid from the coelom onto the body surface via the dorsal pores (Wolf, 1940).

Manwell (1959) however suggests that the method of *L. terrestris* exchanging gases through a boundary of cuticle, epidermis and hypodermis is relatively inefficient, and that this is compensated by the properties of the haemoglobin in the blood system (see later).

*Rate of Respiration*

As the exchange of gases takes place across the body wall it is to be expected that the respiratory rate will be affected by the size of the animal, and thus by the surface area. Other parameters such as temperature will also affect the respiratory rate (Krüger, 1952, on *E. foetida*).

*L. terrestris*, weighing between 2·5 and 5 g, respires at an average rate of 45·2 mm$^3$ $O_2$/g/hr for the first hour at normal oxygen tensions, this rate gradually dropping to 38·7 mm$^3$ $O_2$/g/hr during a second hour at a temperature of 10 °C (Johnson, 1942). Values of a similar order, ranging between 25 and 105 mm$^3$ $O_2$/g/hr in air and immersed in water were obtained by Raffy. At a temperature of 16–17 °C an individual weighing 1·47 g had a respiratory rate of 70 mm$^3$ $O_2$/g/hr, while another worm of 5·43 g respired at a rate of 31 mm$^3$ $O_2$/g/hr; thus the larger and heavier the animal, and consequently the smaller the surface area:volume ratio the lower the oxygen uptake per gramme of body tissue (Raffy, 1930).

When earthworms are submerged in water the respiratory rate depends directly on the partial pressure of the oxygen dissolved in the water. Table 11 shows that when the partial pressure of oxygen in water drops so does the oxygen uptake rate of the earthworm, and when the partial pressure rises so does oxygen consumption (Raffy, 1930).

The rate of oxygen consumption is also greatly affected by the ambient temperature, a parameter likely to be of more importance to an earthworm in nature than the partial pressure of oxygen in water. Pomerat and Zarrow (1936) observed a variation between 25 and 250 mm$^3$ per individual per 30 minutes in a temperature range from 9–27 °C using the Warburg method to detect respiratory exchange. They found no difference existed in the level of respiration between normal, decerebrate and animals with no sub-oesophageal ganglion.

The problems of acclimatization which have received much attention recently in many groups have been studied in the species

*Lumbriculus* and *Eisenia foetida* by Kirberger (1953). She finds that although $Q_{10}$ ( = rise in respiratory rate for 10 °C rise in temperature) values may change only slightly the respiration curves of all cold acclimatized animals shifts to the left as compared with warm acclimatized animals. That is to say at any given temperature animals maintained previously at a low temperature respire more slowly than individuals from a warmer position. Thus specimens from the tropics respire faster than specimens from colder climes when at the same temperature.

As we have seen above when the temperature rises *L. terrestris*

TABLE 11

EFFECT OF PARTIAL PRESSURE OF OXYGEN ON RESPIRATION OF *Lumbricus* OF TWO SIZES
(From Raffy, 1930)

| Wt. of animals g | p. p. $O_2$ ml/l. $H_2O$ | $O_2$ uptake mm$^3$/g/hr |  |
|---|---|---|---|
| 5·14 | 5·48 | 49 |  |
|  | 4·52 | 32 | at 17 °C |
|  | 4·04 | 20 |  |
| 1·99 | 6·635 | 146 |  |
|  | 8·289 | 173 | at 21 °C |
|  | 18·410 | 180 |  |

respires more quickly. Earthworms from the tropical lands, however, normally live under temperature conditions higher than earthworms in temperate regions. And as such they are found to respire at higher rates. Table 12 gives the rates of oxygen consumption of four species of tropical earthworms at various temperatures. Comparison with the rates exhibited by temperate species, however, reveals that at the same temperature the rates of both types are similar. The higher respiratory uptake of tropical species is a reflection of the higher environmental temperatures that normally surround them. As is true for temperate species individual respiration rates depend upon the size and volume of the animal, smaller individuals respiring faster than large ones, and *Glossoscolex*, a larger species, respires at a lower rate than either of the smaller species *Pontoscolex* or *Pheretima* (Mendes and Valente, 1953).

Variations in respiratory rate which are correlated with body size are also mentioned by Saroja (1959) working on *Megascolex mauritii*. Respiration of this species rises with temperature over a range of 15–35 °C, but the relative rate of increase was greater for small worms (0·15 g) than for large individuals (1·5 g). $Q_{10}$ also shows increases with decreasing weight but no systematic variation

TABLE 12

RATE OF RESPIRATION, BEFORE AND AFTER STARVATION, OF FOUR TYPICAL OLIGOCHAETES

| Species | Temp. (°C) | Wt. in g | Rate fresh wt. mm³/g/hr | Rate after 24 hours starvation | Author |
|---|---|---|---|---|---|
| *Pontoscolex* sp. | 25 | 0·36–1·05 | 130–284 | 145–272 | Mendes and Valente (1953) |
| *Pheretima hawayana* | 25 | 0·57–2·017 | 110–230 | 61–271 | |
| *Glossoscolex* sp. | 25 | 6·67–18·884 | 39–100 | 38–109 | |
| *Megascolex mauritii* | 15 | 0·15–1·5 | 58–86·7 | | Saroja (1959) |
| | 25 | ,, | 100–233 | | |
| | 35 | ,, | 149–400 | | |

of $Q_{10}$ with temperature is found. *Megascolex mauritii* normally lives in a temperature range of 25–30 °C and over this range oxygen uptake shows a linear proportionality to surface area, but at temperatures below 35 °C the ratio oxygen consumption: weight shows a consistent decrease with the rising temperature (Saroja, 1959).

Gaseous exchange in these tropical earthworms takes place across the body surface as it does in *Lumbricus* via the intraepidermal capillaries. The respiratory rates of *Pontoscolex*, *Pheretima* and *Glossoscolex* all rise when adrenaline ($10^{-1}$ g/l.) is applied to the surface of the body. This is possibly due to dilation of the skin capillaries, and a reverse effect is obtained upon the use of acetylcholine, the respiratory rate being decreased, probably due to constriction of the capillaries. These results may be spurious as very concentrated adrenaline and acetylcholine solutions were used in order to overcome the barrier provided by the mucus secretion of the body wall. The mucus does not contain any cholinesterase activity (Mendes and Nanato, 1957). It would be interesting to know if respiratory rates are increased when adrenaline is injected into the blood system. The pseudohearts and general circulation of *L. terrestris* function more rapidly when adrenaline ($10^{-8}$ g/l.) is injected (Prosser and Zimmerman, 1943), and this may be reflected by the oxygen consumption of the animal.

*Patterns of Gaseous Exchange*

That the respiration of animals, and indeed plants as well, is not a constant throughout the day, or throughout the seasons of the year, is now well known and has recently been reviewed by Harker (1958). Diurnal, lunar, monthly and annual cyclic changes of respiratory rates are known to be widespread.

The earthworm also shows a fluctuating respiratory rhythm. Records continuous for up to 2 months are available of the course of respiration of *L. terrestris*. Experiments were carried out in unvarying conditions of temperature, humidity and low illumination using the buoyant respirometer described by Brown (1954). Analysis of kymograph recordings made in this way led Ralph (1957) to conclude that earthworms show a diurnal rhythm in oxygen consumption which showed maximal rates at about 6 a.m. and again at approximately 7 p.m. Minimal rates occurred at about 10 p.m. Further statistical manipulation of the results was claimed to show that lunar day (24·8 hr) and lunar (29·5 days) cycles were also part of the make up of respiratory rhythms in *L. terrestris* (Fig. 27).

The method used in this study has been severely criticized by Cole (1957) and further investigations with a different method might be advisable. This is particularly pertinent in view of the fact that Ralph (1957) found that periods of maximum locomotor activity are associated with low oxygen consumption rates (Fig. 27) and suggested that earthworms accumulate oxygen debts at such times. Oxygen debts are known to occur under experimental conditions (Davis and Slater, 1928) but there is no evidence that similar conditions occur in the field. On the other hand neurosecretory

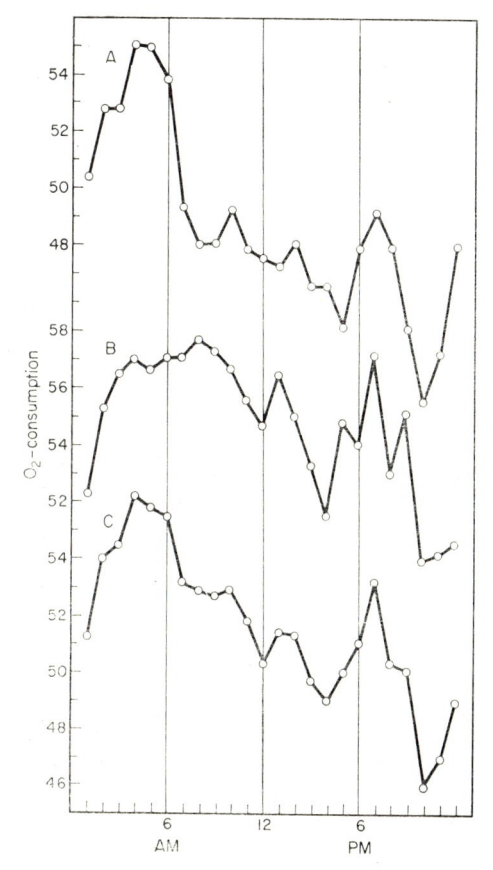

FIG. 27. Diurnal rhythm of $O_2$ consumption in *L. terrestris* for two consecutive semilunar periods. A. May 2–16, B. May 17–31 and for the entire lunar period, C. May 2–31, 1954 (from Ralph, 1956. Copyright University of Chicago).

cells may influence locomotor activity, as in cockroaches (Harker, 1960) and thus lead to a discrepancy in locomotor movements and oxygen consumption. It seems an anomaly however, that a period of active movement should be a low respiratory period, and vice versa, particularly when such periods last for hours rather than a few minutes.

*Effect of Variations in Oxygen Availability*

When earthworms are living in normal air conditions, as when feeding on the surface of the ground, respiration progresses at an average rate of 38·7–45·2 mm³/g/hr. The partial pressure of oxygen is approximately 152 mm Hg in normal atmospheric air. Earthworms, however, spend long periods underground in burrow systems which may be deep, down to 6 feet, and narrow. It is conceivable, but unlikely, that the air in these channels becomes depleted of oxygen as it is utilized by the inhabitant of the burrow,

TABLE 13

RATE OF RESPIRATION OF *Lumbricus terrestris* AS A FUNCTION OF THE PARTIAL PRESSURE OF OXYGEN IN THE ATMOSPHERE
(From Johnson, 1942)

| $O_2$ p.p. mm Hg. | 152 | 76 | 38 | 19 | 8 |
|---|---|---|---|---|---|
| Rate mm³/g/hr | 38·7–45·2 | 35·2–42.5 | 23·3–25·0 | 15·4–16·7 | 6·7–7·2 |

but oxygen tensions may fall considerably without the activity of worms (Russell, 1950). Laboratory experiments have indicated that oxygen tensions must be quite drastically lowered to affect the rate of respiration of earthworms. A decrease of 50% in the partial pressure of oxygen (76 mm Hg) does not alter oxygen consumption rates but a reduction to 25% (38 mm Hg) depresses respiration by 55–60%, and conditions of greater oxygen lack decrease the rate further still (Table 13) (Johnson, 1942).

The partial pressure of oxygen also affects the respiratory rate of *Lumbricus* when immersed in water as mentioned previously (p. 87). The fresh-water oligochaete *Tubifex tubifex*, however, appears unaffected by changes in oxygen availability. The worms live in a mud substratum, which may at times become almost devoid of oxygen but the rear end of the body projects into the water flowing above them. Fox and Taylor (1955) found that adult *Tubifex* survive equally well in water containing 21% oxygen i.e. air saturated, and in water with only 4% oxygen. Conditions such as these may be expected to occur in situations suitable as habitats for these worms. In the artificial conditions of complete saturation of the water with 100% oxygen the animals die in a few days. No details are given of the rates of respiration in these cases.

The survival of adult *Tubifex* then is not dependent upon the provision of a high partial pressure of oxygen. The situation with regard to newly hatched *Tubifex*, however, is very different. The hatching of cocoons occurs equally well in water containing 21% oxygen, but after one month more worms were living in 4% oxygenated water than in air saturated water. Moreover the individuals in 4% oxygen were larger, having a mean volume of 5·68 mm$^3$, compared with 1·79 mm$^3$ of those in air saturated water (Fox and Taylor, 1955).

The influence of complete oxygen saturation of water has also been studied by Walker (1959) using pressures above atmospheric. Eight hours treatment with 100% $O_2$ at four atmospheres pressure has no noticeable effect on *Tubifex* but more prolonged exposure causes death, in much the same way as reported by Fox and Taylor (1955). Interruption of the treatment prolongs the survival time, suggesting that some recovery from oxygen damage is possible. The provision of a high osmotic pressure externally in the form of a salt solution promotes survival, as it also does in cases of heat stress, carbon monoxide and hydrogen peroxide poisoning. But the application of salt solutions to a fresh-water dweller must itself have deleterious effects and provoke the occurrence of homeostatic mechanisms to maintain internal equilibria.

Changes in the proportion of the body of adult *Tubifex* is exhibited in response to oxygen variation. The river and lake-bottom mud in which the oligochaetes live is also the home of many bacteria and the activity of the latter may at times reduce oxygen tensions virtually to zero. Such oxygen depletion is followed by extension of the body of *T. tubifex* to a length ten to twelve times normal, thus projecting the rear end away from the substratum into water more likely to contain oxygen, and also presenting a greater surface area for gaseous exchange to take place. At the same time rhythmic corkscrew movements agitate the water surrounding the worms and stimulate currents of water from above containing more oxygen to flow down around the animals. These movements are inversely proportional to the oxygen supply below the water surface (Alsterberg, 1922, Dausend, 1931). Dobson and Satchell (1956) also have interesting speculations on the availability of oxygen with regard to bodily proportions in the earthworm *Eophila oculata*. Prolonged exposure to low oxygen tensions does

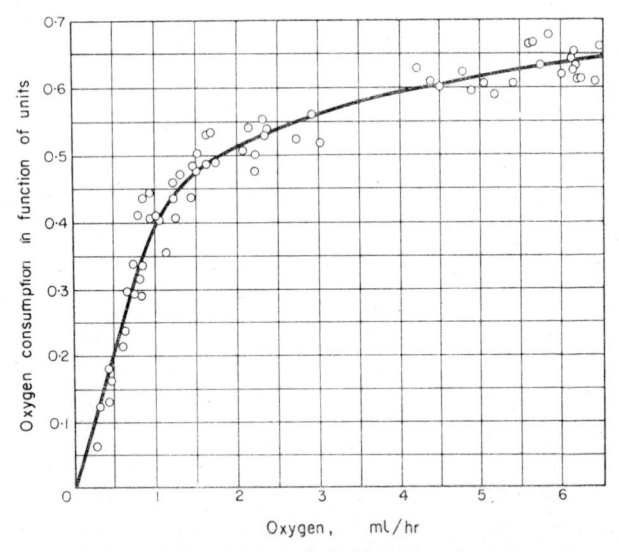

FIG. 28. Oxygen consumption of *Tubifex* as a function of oxygen tension of surrounding water (from Dausend, 1931).

not, however, lead to an increase in the amounts of haemoglobin present.

The respiratory rate of *T. tubifex* is virtually constant at all oxygen tensions above 0·5% $O_2$, at which concentration the uptake is 0·6 mm³/g/hr at 19 °C. If the oxygen tension falls further the respiratory uptake falls, slowly at first, to 0·4 mm³/g/hr at 0·1% oxygen, and then decreasing rapidly as the oxygen available approaches 0% (Fig. 28). The haemoglobin of the blood is responsible for only a fixed fraction of the respiratory exchange since carbon monoxide, which combines readily with haemoglobin depresses respiration by about a third at tensions above 0·1% (Dausend, 1931 (Fig. 29).

It seems then that *T. tubifex* is well adapted to utilize the normally low tension of oxygen available at mud surfaces and that it can survive complete air saturation of the water providing the animal has reached maturity. Respiration rates are constant over a wide range of oxygen tensions, falling only when oxygen reaches

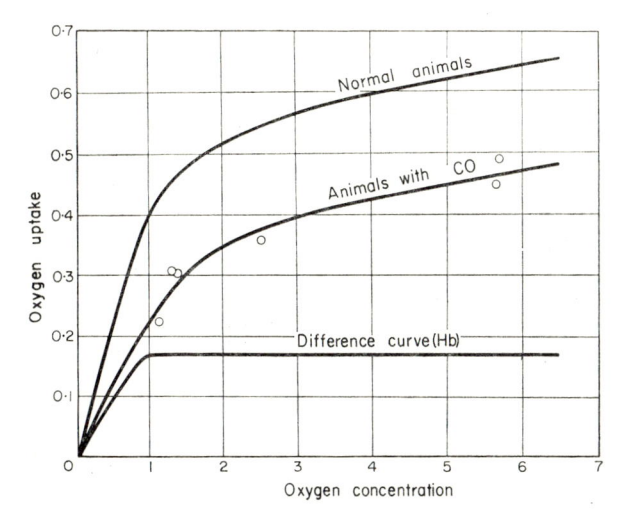

Fig. 29. *Tubifex* oxygen consumption. (a) normal animal, (b) animal treated with carbon monoxide, (c) difference between (a) and (b) representing fraction of oxygen carried by haemoglobin (from Dausend, 1931).

a very low level at which time the haemoglobin of the blood system takes over a greater fraction of the gaseous exchange.

*Effect of Carbon Dioxide*

The respiration of animals is affected not only by the availability of oxygen but also by the presence of large quantities of carbon dioxide in the atmosphere. Little information is available regarding the respiration of earthworms in the presence of heightened concentrations of carbon dioxide though it is generally believed that this product does not alter respiration greatly. Intense activity or prolonged incarceration within the burrow may cause a build up of carbon dioxide in the atmosphere. Shiraishi (1954) placed *E. foetida* in an artificial burrow and found that when $CO_2$ is passed over the animal no effect is noted upon its movement until a concentration of 25% is reached. Any further increase leads to the withdrawal from the area of high concentration. It is possible that such high concentrations may occur, though Russell (1950) says that the maximum concentration of $CO_2$ obtained in soil is only 7%, particularly so after heavy rain has occluded the orifices of the burrow and the well-known migration on the surface of the ground of some species after rain may be ascribable to this reaction. It has also been suggested, however, that this movement is due to lack of oxygen and no conclusion can yet be reached (Ledebur, 1939). Svendsen (1957) suggests that this may simply be a normal dispersal mechanism.

In the case of *T. tubifex* excess carbon dioxide does not accelerate the waving movements in response to lack of oxygen, on the contrary it may depress and stop this activity (Alsterberg, 1922, Dausend, 1931).

Thus far we have discussed the mechanisms, pattern and bodily reactions of oligochaetes, in particular *Lumbricus* and *Tubifex*, in respiration. With the exchange of oxygen and carbon dioxide across the body wall now dealt with it remains to consider the methods, if any, used to transport these gases, and how they are involved in cell metabolism.

*Blood Vessels and Haemoglobin*

First transport—there is an extensive closed blood system composed of vessels and capillaries in the oligochaetes the exact layout of which can be obtained from any standard text book such as Borrodaile, Eastham, Potts and Saunders or Parker and Haswell or a more detailed account is in Stephenson (1930). The major vessels, one ventral and one dorsal, run longitudinally through the body. In the ventral vessel blood moves to the posterior extremity whilst the dorsal vessel collects blood back from the segmental vessels and passes it forward. The dorsal vessel pulsates as also do certain commissural vessels anteriorly, the pseudohearts. The frequency with which the vessels contract varies according to the temperature (Rogers and Lewis, 1914) and this is confirmed by Conroy (1960), and can be influenced by injected adrenaline and

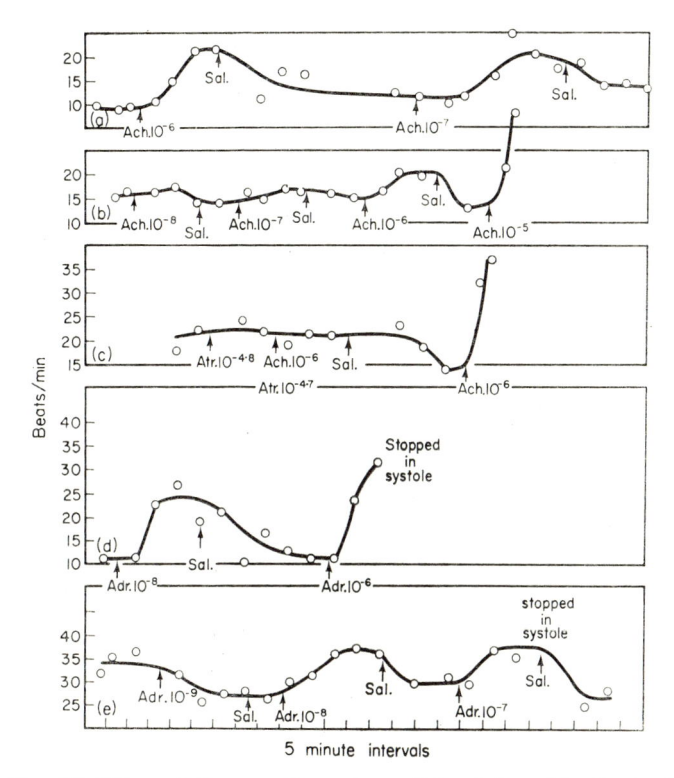

FIG. 30. Effect of drugs on heart beat of *Lumbricus*. (a) effects of acetyl-choline (ACh) $10^{-6}$ and $10^{-7}$ and recovery in saline (Sal.) on one heart in a worm from which the ventral nerve cord was removed from segments 4–18. (b) effect of acetylcholine on dorsal vessel of an earthworm. (c) effect of acetylcholine after atropinization and after washing out the atropine. (d) and (e) effects of adrenaline (Adr.) on two earthworm hearts. Each point represents the average of several measurements (from Prosser and Zimmerman, 1943).

acetylcholine (Prosser and Zimmerman, 1943). The natural occurrence of these substances has recently been conclusively demonstrated in oligochaetes. Muscular movements of the gut and of the body wall also help to move the blood through the vessels but pressure within the vessels is always very low, ranging from 4·4–5·5 mm Hg when the animal *Lumbricus* is at rest up to something more than 9·3 mm Hg when active (Prosser *et al.*, 1950). Extensive blood capillaries are present in both body and gut walls.

The fluid circulating within the vessels is red in colour due to the presence of the respiratory protein haemoglobin. This substance is not contained within corpuscles as in vertebrates but is in physical solution in the plasma. Amoebocytes are present in the blood which carry out a phagocytic action, but are unable to traverse the walls of blood vessels (Tuzet and Attisso, 1955). These cells are not associated with haemoglobin. Not much is known of other blood constituents such as the ions, sugar, amino-acids, etc.

*Haemoglobin*

The function of respiratory pigments is to combine with oxygen at the surfaces available for such a reaction i.e. the body wall of oligochaetes and transfer it to the various tissues where it is released as required in response to a low oxygen tension. It is possible to envisage three methods by which such an action may be carried out.

(1) Transport and release of oxygen by haemoglobin may occur incessantly during life.

(2) The haemoglobin may only come into play when oxygen tensions outside the body fall to such a low level that simple diffusion inwards through the surface is no longer sufficient to keep a high level of dissolved oxygen in solution in the blood plasma circulating round the body, or

(3) The respiratory pigment may act as a temporary store of oxygen to be used only in times of severe oxygen stress (Fox, 1940). The appropriate position of earthworm haemoglobin in the above scheme of things was unsettled until the work of Johnson (1942). The earlier experiments of Jordan and Schwarz (1920) and Dolk and Van der Paauw (1929) suggested that haemoglobin in *Lumbricus* acted only as a store to be used as and when oxygen stress was encountered. Thomas (1935) criticized these conclusions and on the basis of experiments with five worms decided that

TABLE 14

Respiratory Rate of *Lumbricus terrestris* in $MM^3/G/HR$ in Presence and Absence of Carbon Monoxide (From Johnson, 1942)

| O₂ p.p. mm/Hg | 152 | +CO | 76 | +CO | 38 | +CO | 19 | +CO | 8 | +CO |
|---|---|---|---|---|---|---|---|---|---|---|
| Rate in 1st hr. | 45·2 | 43·5 | 42·5 | 46·2 | 25·0 | 28·2 | 16·7 | 17·8 | 7·2 | 7·6 |
|  | — | — | — | — | — | — | — | — | — | — |
| Rate in 2nd hr. | 38·7 | 29·6 | 35·2 | 25·2 | 23·3 | 15·7 | 15·4 | 12·9 | 6·7 | 7·0 |

little if any of the oxygen uptake and transport in *L. terrestris* was carried out by haemoglobin. The affinity of haemoglobin for oxygen is outweighed only by its affinity for carbon monoxide. In combining with carbon monoxide the amount of haemoglobin available for reaction with oxygen is reduced. By adding carbon monoxide to gas mixtures it is possible to investigate the efficiency of haemoglobin systems. Johnson (1942) in fact used mixtures of carbon monoxide and oxygen, having sufficient CO to saturate

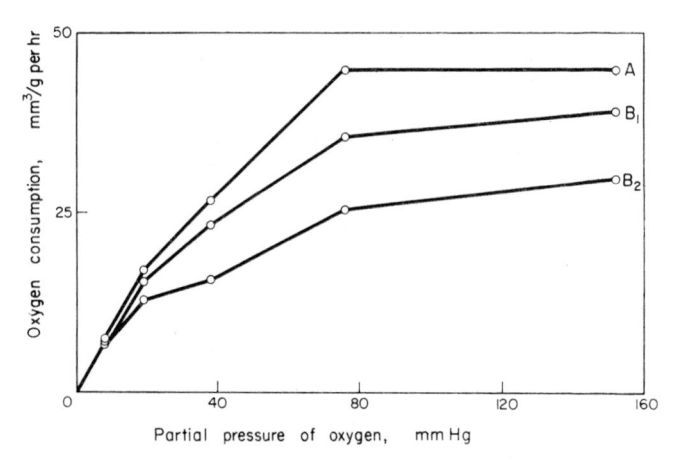

Fig. 31. Mean rate of oxygen consumption of earthworms at 10 °C at different oxygen pressures. A, during first hour in absence of CO; $B_1$ during second hour in absence of CO; $B_2$ during second hour in presence of CO (from Johnson, 1942).

95% of the haemoglobin of the blood of *Lumbricus*. The oxygen consumption of carbon monoxide treated animals is significantly lower than that of untreated animals at all oxygen pressures down to 8 mm Hg but not at this partial pressure (Table 14, Fig. 31).

In order to check that this depression of oxygen consumption was indeed due to inhibition of haemoglobin, Johnson (1942) studied the respiratory rate of tissue slices (of the body wall) in the presence of CO, and found that the oxygen uptake of CO treated slices was 110% that of untreated slices. There was no decrease in isolated tissue respiration so any effect of CO is due to haemoglobin blockage in the intact animal. It is clear, therefore, that haemoglobin in the earthworm functions as a carrier of oxygen at all times in fulfilment of function 1. The actual proportion of the body's needs that are carried in this way, however, differ somewhat with changing oxygen tensions. At 152 mm Hg (normal atmospheric air) haemoglobin carries approximately 20% of the total requirements, at 76 mm Hg the proportion is 35%, at 38 mm Hg 40% and at 19 mm Hg 22%. From these figures Johnson (1942) deduced that the loading pressure of earthworm haemoglobin is higher than 19 mm Hg at 10 °C. The remainder of the earthworm oxygen requirement is carried in physical solution.

The respiration of *Tubifex* is also decreased by carbon monoxide treatment, about a third of the normal oxygen consumption being depressed when the oxygen tension is high, proportionately less at lower tensions (Dausend, 1931).

*Oxygen Dissociation Curves*

Oxygen carried in combination with haemoglobin to the tissues is there released in response to a relatively low oxygen tension. It is possible to obtain curves that indicate the way in which haemoglobin combines and releases oxygen. Only within the last few years have techniques been worked out for preparing such curves using the small quantities of blood available from earthworms. By the use of spectrophotometric methods, however, Haughton, Kerkut and Munday (1958) and Manwell (1959) have been able to elucidate the proportion of reduced and oxygenated haemoglobin in a mixture of the two forms.

The curve obtained plotting percentage saturation of haemoglobin with oxygen, against partial pressure of oxygen available, is sigmoid in shape. The haemoglobin remains 95% saturated until a very low partial pressure is reached (Fig. 32a, Table 15). Therefore at all normal atmospheric levels of oxygen the haemoglobin of earthworm blood is saturated with oxygen. Only when the external tension drops greatly does the haemoglobin unload, this occurring normally at the tissues. The unloading tension, deduced by Johnson (1942) as rather more than 19 mm Hg oxygen at 10 °C is incorrect, as Manwell (1959) working at the same temperature has shown that *Lumbricus* haemoglobin is 85% saturated at this

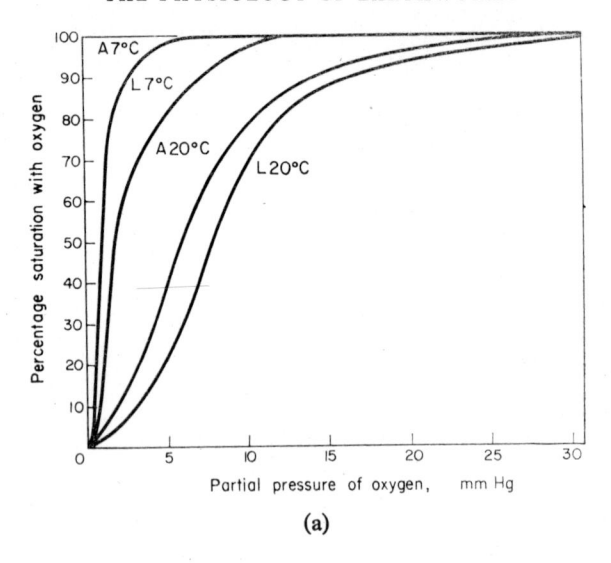

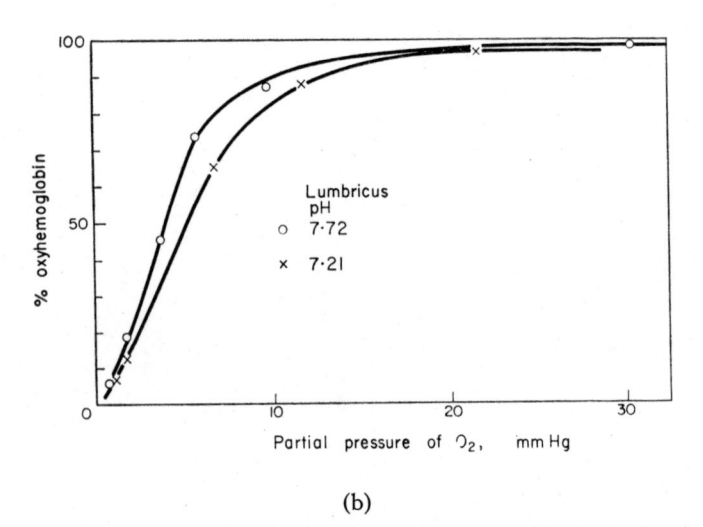

Fig. 32. Oxygen dissociation curves of haemoglobin from *Lumbricus* and *Allolobophora*. (a) shows the effect of temperature on the curves; A. referring to *Allolobophora*, L. to *Lumbricus*. (b) shows the effect of pH on the dissociation curve of *Lumbricus*. (a) from Haughton, Kerkut and Munday, 1958, (b) from Manwell, 1959, (by permission of the Wistar Institute).

partial pressure, 80% saturated at approximately 8 mm Hg of oxygen and 50% saturated at 3·5–4·8 mm Hg of oxygen.

Temperature and pH both affect the position and shape of the dissociation curves. At low temperatures the haemoglobin has a more pronounced affinity for oxygen the curve becoming steeper (Fig. 32a) and the unloading tension lower (Table 15). This is true for two species, *L. terrestris* and *A. longa* (Haughton, Kerkut and Munday, 1958). The dissociation curves of these two species are not identical at the same temperature and it is thought that this may be a reflection of the somewhat different ecology of the two species, one remaining active throughout the year (*Lumbricus*) and the other aestivating in the summer (*Allolobophora*).

TABLE 15

VALUES FOR PARTIAL PRESSURES OF OXYGEN (mm Hg) REQUIRED TO SATURATE THE BLOOD
(From Haughton, Kerkut and Munday, 1958)

|  | 50% saturation ||| 95% saturation |||
| --- | --- | --- | --- | --- | --- | --- |
|  | 7 °C | 10 °C | 20 °C | 7 °C | 10 °C | 20 °C |
| *L. terrestris* | 2 | 3·5–4·8 | 8 | 9 | *ca.* 18 | 22·5 |
| *A. longa* | 0·7 |  | 6 | 3·75 |  | 17·5 |

A change in pH from 7·72–7·21 displaces the curve to the right (Fig. 32b), that is to say there is a "Bohr" effect, an increasing acidity leading to the raising of the unloading tension (Manwell, 1959). Thus if carbon dioxide accumulated in the blood and lowered the pH the unloading tension would rise and the respiratory rate would also rise. As mentioned previously, however, carbon dioxide in the atmosphere has no effect on respiration rates, suggesting that the mechanism of acid–base balance effected most likely by the calciferous glands, is efficient in binding excess $CO_2$ rapidly.

Manwell (1959) suggests that these curves indicate that the haemoglobin of *Lumbricus* is suitable for unloading oxygen at sites of low internal oxygen tensions, but would not be of great significance at low atmospheric tensions when there would be an insufficient gradient to load the haemoglobin. In such a case it would be expected that carbon monoxide would have a relatively greater inhibitory effect on respiration at high rather than low oxygen tensions as shown by Johnson (1942). Considerable amounts of oxygen are carried in physical solution at high oxygen tensions, however, and the effect of carbon monoxide may be diminished by this fact. This has been demonstrated by Krüger and Becker (1940).

Data for the aquatic oligochaete *Tubifex* indicates that 50% saturation is reached at 0·6 mm Hg oxygen thus making a steep concentration gradient across the body wall.

The function of haemoglobin as a carrier of oxygen is also confirmed for three tropical species, *Protoscolex*, *Glossoscolex* and *Pheretima hawayana*, the respiration of which is depressed by about 50% when the atmosphere contains 20% carbon monoxide (Mendes and Valente, 1953).

*Chemical and Physical Properties of Oligochaete Haemoglobin*

The haemoglobin molecule of *L. terrestris* is a large one, with an estimated molecular weight of 2,946,000, having a sedimentation rate in the ultracentrifuge of $60·8 \text{ mm}^3 \times 10^{-13}/\text{sec/dyne}$ at 20 °C. On the basis of the porphyrin ring structure of haemoglobin the molecule represented in *Lumbricus* blood contains 144 iron atoms in some seventy-two units of 34,500 mol. wt. The isoelectric point of the protein is pH 5·28 and upon hydrolysis this breaks down to give at least six amino acids (see p. 5), Svedberg (1933).

Haemoglobin can be sedimented from blood plasma by ultracentrifugation at 76,000 *g*. The material is soluble in water and on standing for one week at 0 °C and pH 7 it turns brown with the appearance of an absorption peak at 645 m$\mu$, corresponding to the formation of ferrihaemoglobin which disappears on adding $Na_2SO_4$. Two spectral bands are usually shown by solutions of haemoglobin (Table 16).

The reduction of the oxy-form of haemoglobin causes a slight shift in the absorption spectrum as shown for *Pheretima* in Table 16. The time taken for the haemoglobin to unload half of its

TABLE 16

SPECTRAL ABSORPTION OF OLIGOCHAETE HAEMOGLOBINS

|  | Oxyhaemoglobin || Reduced haemoglobin || Met-haemoglobin | Carboxy haemoglobin || Author |
|---|---|---|---|---|---|---|---|---|
|  | α | β | α | β |  | α | β |  |
| *Lumbricus* | 576 | 544 |  |  | 645 | 570 | 535 | Salomon, 1941 |
| *Pheretima* | 574 | 538·1 | 567 | 549 |  |  |  | Kobayashi, 1935 |

charge of oxygen is 0·070 sec at 23 °C and pH 8 (Salomon, 1941).

Information gained from centrifugation and from the spectral analysis leads to the conclusion that one type of haemoglobin is present. The technique of alkaline denaturation, however, can be used to determine whether one or more types, having essentially similar properties, are to be found. Blood obtained fresh and brought to an alkaline pH (10·5–13) is converted from oxy-haemoglobin to alkaline globin haemochromogen. Under controlled conditions this reaction can be followed by monitoring the proportion of unchanged oxyhaemoglobin in solution.

The time taken for the reaction to go to completion differs from species to species, and from pH to pH. At pH 12·7 complete denaturation takes 23 minutes for *L. terrestris* but only 6 minutes for *A. longa*. *L. festivus* 11·5 minutes, *E. foetida* 5 minutes and *L. rubellus* 2·5 minutes, all show individual denaturation times (Haughton, Kerkut and Munday, 1958). A denaturation time of about 24 minutes is given by Manwell (1959) for *L. terrestris* haemoglobin at pH 11·7. Denaturation proceeds more slowly in less concentrated alkali, taking about 34 minutes at pH 11·2 and 50 minutes at pH 10·92 (Manwell, 1959).

The construction of curves showing the course of denaturation reveals that the process is not a simple linear event, but is composed of a number of phases. According to Haughton, Kerkut and Munday (1958) the process takes place in two stages, the initial stages of denaturation proceeding more quickly than the later stages. The slope of the curve is steep and straight until the concentration of unchanged oxyhaemoglobin is only 30% and then the line breaks to give a different more gentle slope until completely denatured (Fig. 33). Arguing from the different denaturation times of mammalian adult and foetal haemoglobins (Brinkman and Jonxis, 1937) it is thought that the two portions of the curve are indicative of two forms of haemoglobin in earthworms. Very few points are given for the initial portion of the process, however, and more information is given by Manwell (1959). He finds evidence that the denaturation process is triphasic rather than biphasic. Three equal stages occur, each accounting for about a third of the haemoglobin. The form of the denaturation is constant whatever the pH used. Instead of postulating three types of

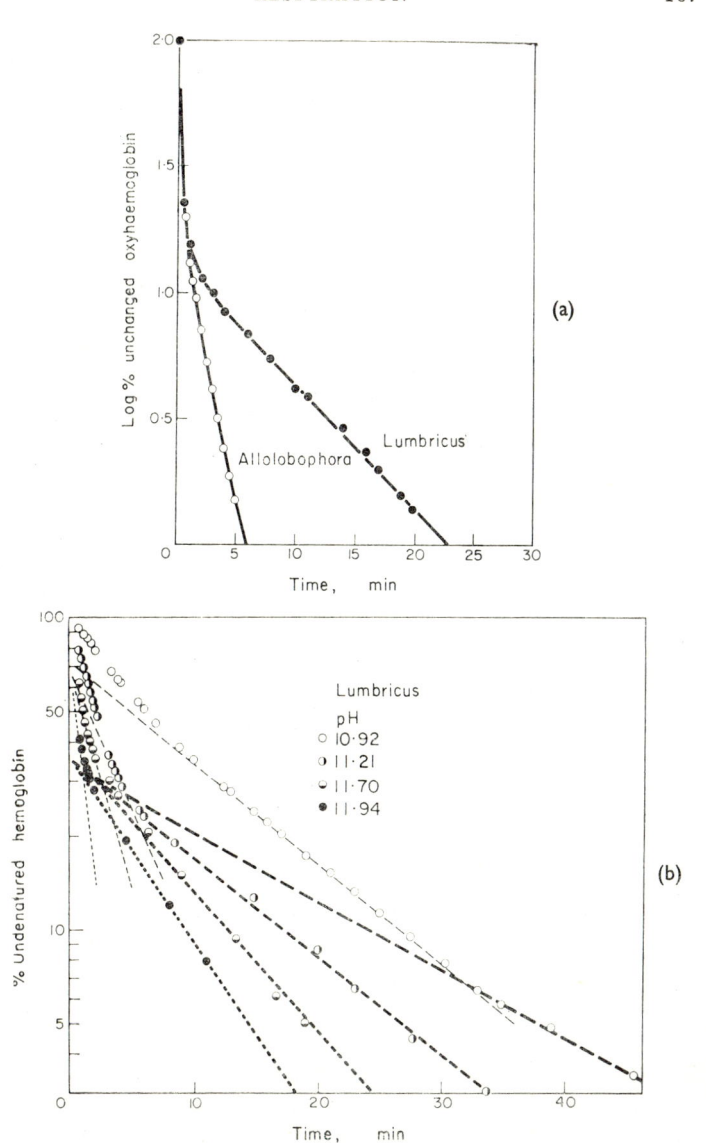

FIG. 33. The course of alkaline denaturation of haemoglobin from *Allolobophora* and *Lumbricus*; (a) at pH 12 (Haughton, Kerkut and Munday, 1958); (b) a nest of curves for pH 10·92, 11·21, 11·70 and 11·94 (Manwell, 1959).

haemoglobin Manwell (1959) suggests that the triphasic denaturation represents a three-step denaturation of one haemoglobin type that has three monomolecular forms. This idea would support the sedimentation and absorption spectra data, although if more than one haemoglobin is present it is not inconceivable that the molecular weights and hence sedimentation constants will be the same.

A little more information is given by electrophoretic studies on the blood of *L. terrestris*. A spot of blood, separated by electrophoresis on a cellulose acetate membrane, gives rise to a number of fractions. At least three are observed, and one appears only on staining with nigrosin. The other two bands are both visible to the naked eye as yellow-orange in colour and are obviously the haemoglobin of the blood. They stain with leucomalachite green, commonly used for the demonstration of haemoglobin. Similar results have been obtained for *A. longa*, and *E. foetida* (Laverack, 1960, unpublished). The evidence supports the theory that two haemoglobins are actually present at the same time, and it is felt that more information could be gained from experiments using moving boundary electrophoresis.

*Cellular Respiration*

Ultimately the process of respiration involves a consideration of the energy cycles of the cells that make up the body of the animal. The oxidation–reduction processes of the cells are the final links in the chain that started with the exchange of gases across the body wall. These oxidation processes of the cells of earthworms involve a heavy metal prosthetic group oxidase, presumably a cytochrome–cytochrome oxidase system. This series of enzymes is blocked by KCN, and Petrucci (1955) finds that the respiratory rate of *E. foetida* in the presence of KCN is only 9% of the normal level. *Peloscolex velutinus* has a KCN resistant respiratory fraction amounting to 16%. Thus the heavy metal oxidase accounts for 84% of cellular respiration in *P. velutinus* and 91% in *E. foetida*. The concentration of cyanides required to block 50% of the oxygen uptake is $7.3 \times 10^{-5}$ M KCN for *E. foetida* and $11.8 \times 10^{-5}$ M KCN in the case of *P. velutinus*.

Carbon monoxide, which combines with haemoglobin to the exclusion of oxygen, depresses the respiratory exchange of *P. velutinus* and *E. foetida*. It reduces respiration by 50% when

$CO:O_2$ is in the ratio 16:1 for *E. foetida* and 80:1 for *P. velutinus* with respect to cyanide sensitive systems.

Estimates of the cytochrome oxidase activity in tissue homogenates reveals that *E. foetida* contains more than *Peloscolex*. Homogenates from *E. foetida* have an oxygen consumption the equivalent of 13·5 times the rate of the intact animal per g. Minced tissues from *Peloscolex* respire only 2·5 times as fast as the normal whole animal. The ratio of $CO:O_2$ that causes 50% reduction in respiratory rates is now only 9·2:1 for *Eisenia* and 21:6 for *Peloscolex*. The inhibition of cytochrome oxidase by carbon monoxide is almost completely suppressed by illumination of preparations with blue light, unless haemoglobin is present in the homogenates of *Peloscolex* (Petrucci, 1955).

## Intermediate Metabolism

The metabolic cycles and mechanisms of vertebrate tissues are by now known. The series of reactions that go to make up the tricarboxylic acid cycle, anaerobic glycolysis, the enzymes involved and the end products are well known. This is not the case with invertebrate animals in general and the oligochaetes in particular. Very little experimental evidence is available and although the indications are that basically the system is as for vertebrates this must be subject to revision as further work progresses.

*Anaerobic Metabolism and Glycolysis*

It is reasonably certain that special mechanisms have evolved in the oligochaete metabolism. For example *Tubifex*, living in mud, is able to survive in the complete absence of oxygen for up to 60 hours. When air again becomes available the animal respires at a greater rate than normal (Alsterberg, 1922). This must mean that food substances are metabolized, possibly by glycolysis, metabolic products accumulate, and these are oxidized when oxygen again becomes available. Although it may be assumed that lactic acid is formed and oxidized, there is as yet no direct evidence, and no one has shown what tissue levels are reached in prolonged anaerobiosis. It is known that glycogen disappears four times as rapidly in anaerobic conditions as in the presence of oxygen, and is resynthesized when oxygen becomes available again, but no evidence exists to demonstrate the course the breakdown and synthesis take (Alsterberg, 1922).

A similar paucity of results exist for *L. terrestris*. An oxygen debt is built up by that species under anaerobic conditions with a concurrent formation of lactic acid. Upon re-admittance of oxygen part of the lactic acid is oxidized, and the remainder resynthesized to glycogen (Davis and Slater, 1928). Glycogen evidently acts as an energy store for it is mobilized from its site in the chloragogen cells and decreases in amount when earthworms are starved (van Gansen, 1956). Glucose-1-phosphate, the first intermediate formed during glycolysis, has been found but there has been no sign of glucose-6-phosphate or later products in the system. However, although glucose-6-phosphate has not been shown chromatographically, de-Ley and Vercruysse (1955) obtained evidence that dehydrogenase systems acting on glucose-6-phosphate, and gluconate-6-phosphate are both present in *Tubifex tubifex* and *L. terrestris* suggesting that the hexose-monophosphate route is extant in oligochaetes. An acid phosphatase enzyme occurs in chloragogen cells but its role is not yet clear (see p. 57).

*Aerobic Metabolism*

It is also as yet too early to state that the Krebs tricarboxylic acid cycle acts in the oxidative metabolism of oligochaetes, although the evidence available certainly suggests that it does.

Some of the intermediate substances in the cycle, and some of the associated enzyme systems involved in the transformation occur in *Peloscolex velutinus* and *E. foetida* (Petrucci, 1952, 1954).

Pyruvic acid, the end product of glycolysis which is injected into the citric acid cycle, and α-ketoglutaric acid, an intermediate substance formed during the conversion of pyruvic acid, are both found in fairly large quantities in the body of these two species (Table 17).

When arsenite is applied to intact worms the oxygen consumption rate falls, indicating an interference with the oxidative metabolism. At the same time the pyruvic acid content of the tissues rises by ten-fold, and α-ketoglutaric acid by 14–25%. This could be due to inhibition of ketoglutaric acid decarboxylase in the Krebs cycle, causing a piling up of α-ketoglutaric acid. Iodoacetic acid, often used as an inhibitor in a stage in the glycolysis of vertebrates, on the other hand, depresses the amount of α-ketoglutaric acid present, pyruvic acid is 200% of the normal value, suggesting an inhibition of the mechanism by which pyruvate is metabolized into the early stages of the citric acid cycle. The inhibition of triosephosphate dehydrogenase (Hawke, Oser, and Summerson, 1954) evidently does not occur here since this would lead to a depression in the amounts of pyruvic acid rather than an accumulation.

The citric acid cycle and the ability of tissue to metabolize the intermediate substances produced is also studied by observing whether or not the addition of these substrates to tissue preparations is followed by an increase in the oxygen uptake of the tissue. Petrucci (1954) added succinate to homogenates of *P. velutinus*

TABLE 17

LEVELS OF PYRUVIC AND α-KETOGLUTARIC ACID IN THE TISSUES OF *P. velutinus* AND *E. foetida* (IN µg/g FRESH WEIGHT)
(From Petrucci, 1952)

| Species | Pyruvic acid | α-ketoglutaric |
|---|---|---|
| *P. velutinus* | 37·1±5·4 | 49·6±11·9 |
| *E. foetida* | 20·44±2·9 | 64·3±23·4 |

tissues and observed that a large increase in the oxygen consumption occurs, as also happens when pyruvic and α-ketoglutaric acid is provided. Mechanisms for the metabolism of the substances do, therefore, exist. Arsenite as mentioned before appears to block the citric acid cycle since the addition of pyruvic acid causes respiratory increase, but α-ketoglutaric acid does not. Somewhere in the stages between these compounds a blockage occurs. Malonate, which inhibits succinodehydrogenase in vertebrates, causes only a 12% drop in oxygen uptake, but completely depresses the increase normally noted on addition of succinate. The comparatively small depression of tissues untreated with succinate, due to malonate, is explained by the endogenous substrate (succinate) being normally present in only small quantities (Fig. 34).

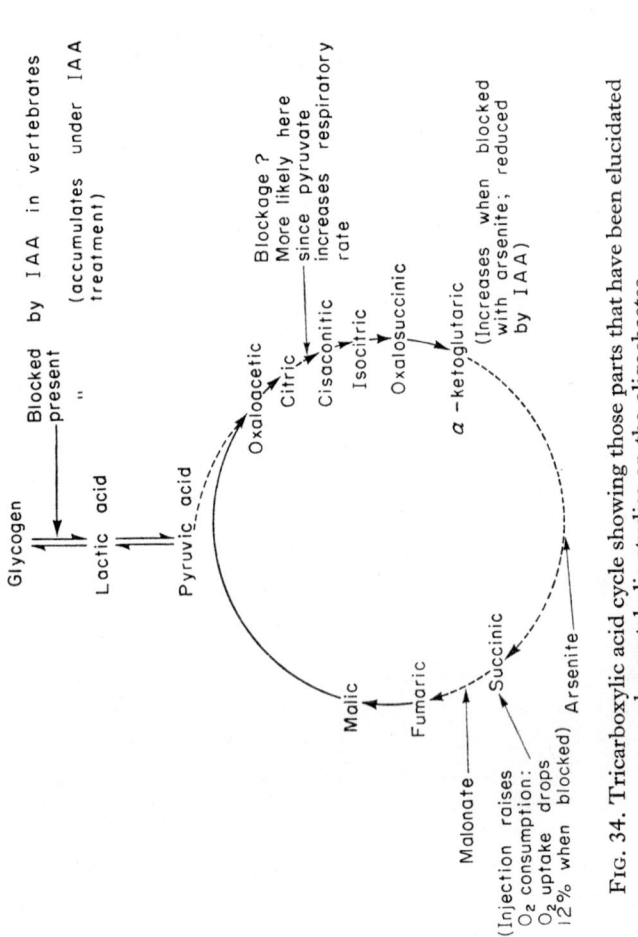

FIG. 34. Tricarboxylic acid cycle showing those parts that have been elucidated by metabolic studies on the oligochaetes.

Thus although the Krebs cycle probably does function in oligochaete tissues as in many other organisms, its final acceptance must be tinged with caution until further steps have been elucidated.

Indications of the course of the energy metabolism of earthworms have also been given by a study of the chloragocytes (van Gansen, 1958). These cells lying on the coelomic surface of the intestine contain large amounts of glycogen (68–325 $\mu$g/mg wt.), amounting to about 7% of the dry weight of the cell, whilst glucose accounts for a further 3·5% of the dry weight. If the chloragogen cells are washed into a test tube with a saline solution and allowed to stand glycolysis occurs and 60–70% of the initial glycogen content disappears. Glucose-1-phosphate appears as an intermediate product in this reaction, but no sign has been seen of glucose-6-phosphate which occurs in vertebrate glycolysis. The enzymatic splitting of glucose-1-phosphate has been accomplished using tissue homogenates suggesting the presence of an acid phosphatase, confirmed by histochemical methods (van Gansen, 1958), and a phosphorylase. Experiments designed to give information regarding the aerobic Krebs cycle, however, were not very successful. Tests for dehydrogenase activity using tetrazolium salts and Thunberg techniques were negative for succinic acid, pyruvate, ethanol, malic acid, citric acid and glycerophosphate, all of which are involved directly or indirectly in the glycolytic or citric acid cycles. This contrasts with the results of Petrucci (1954) outlined above and further work is required to determine what exactly happens, and to demonstrate the modifications, if any, that occur in different cell types.

*The Energy Store*

During the course of intermediate metabolism much energy is released, amounting to at least thirty-eight high energy phosphate bonds in vertebrates when glucose is metabolized to carbon dioxide and water. These high energy bonds are stored in vertebrates as adenosine triphosphate (ATP) and creatine phosphate (CP). When ATP is reduced to ADP it can again regain a high energy bond from CP via the Lohmann reaction. CP is known as a phosphagen.

For many years the invertebrate counterpart of CP was thought to be arginine phosphate (AP), and that this substance was the phosphagen for all invertebrates, including annelids, to the exclusion of the vertebrate counterpart creatine phosphate, save in the echinoderms where both materials occur (Baldwin, 1949). The methods used in the early investigations, however, were not strictly specific and more improved techniques have recently shown that more than one guanidine derivative acting as a phosphagen can be obtained from invertebrates. Among the annelids taurocyamine has been found in *Arenicola*, glycocyamine in *Nereis* and lombricine in *Lumbricus* (Thoai and Robin, 1954).

Earthworm muscles hydrolysed in 6N HCl for 8 hours at 110 °C break down into their constituent parts, and among these Thoai and Robin (1954) found the amino-acid serine and a number of related guanido compounds, formed presumably in combination with arginine. These guanido compounds included guanido-ethylseryl-phosphoric di-ester which they named lombricine. The substance occurs alone in the body wall muscles, but in conjunction with arginine in the alimentary canal musculature. The associated phosphagen, phospho-guanido-ethyl-seryl-phosphate is also obtainable from the same tissues. This is evidently the more important energy store because of its widespread distribution in the tissues, particularly the muscles from which AP is absent.

Since the initial identification of lombricine by Thoai and Robin (1954) the structure, properties and bio-synthesis of lombricine have been elucidated by Ennor and his colleagues (1958, 1959, 1960) and by Pant (1959).

Pure crystalline lombricine has been isolated from earthworms by means of ion exchange, column and paper chromatography (Rosenberg and Ennor, 1959, Pant, 1959). At the same time Rosenberg and Ennor (1959) found serine di-ethyl phosphate (SEP) in tissue extracts, and postulated for it the role of biological precursor to lombricine, since it is present in only small amounts and by non-enzymic guanylation gives rise to a product indistinguishable from natural lombricine. In the biological material the amidine group of lombricine may come originally from arginine by way of transamidination (Fig. 35).

A surprising feature of the chemistry of lombricine was uncovered by Beatty, Magrath and Ennor (1959). Among the aminoacids occurring naturally in living tissues only the L-enantiomorph

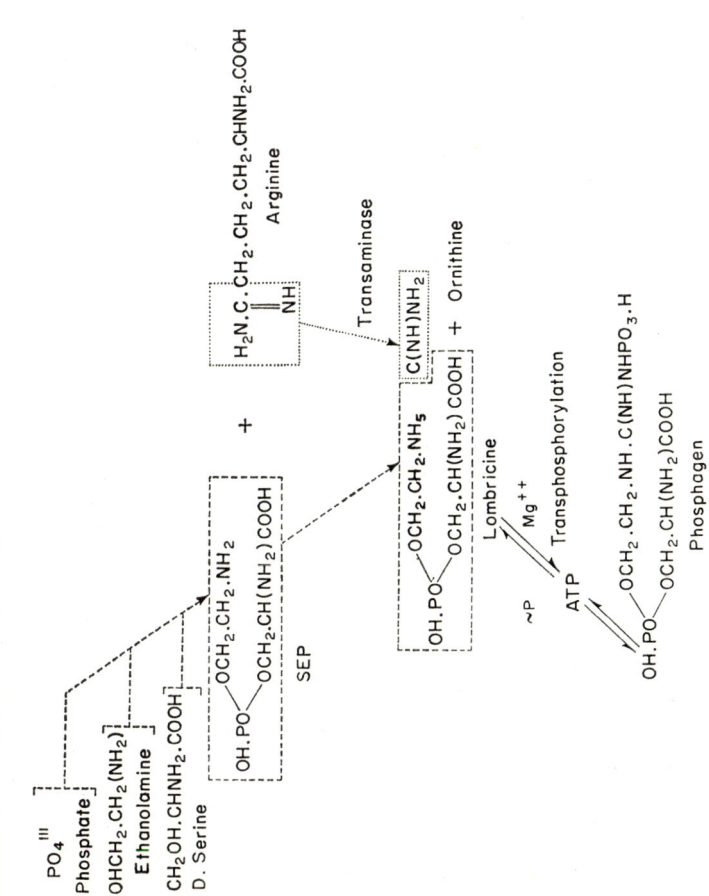

FIG. 35. Metabolic interactions of lombricine, its associated phosphagen and precursors.

is found, but by synthesizing lombricine from serine it is noted that the D form is present in this molecule. D-amino-acid oxidase is without effect on natural lombricine, DL-lombricine and synthetic amino-ethyl-seryl-phospho-di-ester prepared from L-serine. This appears to be the first authentic report of a D-amino-acid to be found in animal tissues.

SEP has also been prepared pure from earthworms and contains serine possessing the D configuration. This is supporting evidence for the suggestion that it is the immediate precursor of lombricine (Ennor et al., 1960).

The biosynthesis of lombricine in *Megascolida cameroni* has been elucidated using radioactive labelled materials. It is found that part of the molecule is derived from arginine, part from ethanolamine and part from serine. L-amidino-$^{14}$C arginine included in the diet of *M. cameroni* leads to a high activity in the guanidinoethanol moeity of lombricine whilst SEP remains almost unaffected, indicating that at least one carbon atom in lombricine originates from arginine in the diet. The inclusion of $^{32}$P, $^{14}$C ethanol-amine or $^{14}$C serine in the food is followed by greater radioactivity in SEP than in lombricine. This also suggests that SEP is indeed the precursor of lombricine and that SEP combines with the amidine group from arginine, as a result of a transamidinase reaction, to give rise to lombricine (Rossitter, et al., 1960).

The probable pathways by which lombricine is formed as indicated above is shown in Fig. 35. If this scheme is correct ornithine is produced as a by-product. This substance has not yet been shown to occur in the tissues other than the chloragogen cells where it is probably an intermediate product in the excretory cycle (van Gansen, 1958). In the muscles it is probably present only in trace amounts, and removed by detoxication mechanisms very swiftly thus being rendered unamenable to analysis.

The associated phosphagen, lombricine phosphate, is formed when high energy phosphate bonds are accepted from ATP. Transphosphorylation from ATP to lombricine is achieved with acetone powders or dialysed enzyme preparations and the maximal enzyme activity is only obtained when Mg ions are present as a co-factor in solution. Activity is gradually lost in dialysed preparations, possibly because of the loss of labile co-factors normally present. The ability of lombricine, in common with taurocyamine and glycocyamine to accept $\sim$P qualifies it as a good candidate for the role of phosphagen in a similar manner to the part played by creatine phosphate in vertebrates, and in contra-distinction to the long accepted role of arginine phosphate in invertebrates. It is interesting to note that arginine is not able to act as a phosphate acceptor using *Lumbricus* enzyme preparations (Pant, 1960).

*Summary*

Respiratory exchange of gases occurs across the body wall and in some species behaviour is modified according to availability of ambient oxygen. Body size has an effect upon the respiratory uptake. Diffusion into the capillaries of the body wall is probably important as gaseous exchange can be modified by various drugs. Haemoglobin in the blood system is in solution and is engaged in uptake and transport at all times. It shows a slight "Bohr" effect, and more than one haemoglobin type is probably present. A certain amount of information is available regarding metabolic processes in the cells of earthworms; the system seems to parallel that of vertebrates although the phosphagen present, lombricine, is a recent discovery.

CHAPTER VIII

# THE PHYSIOLOGY OF REGENERATION

ONE of the most interesting properties of the annelids as a phylum is that of regeneration. Considerable portions of the body can be replaced after loss by injury. The literature on the morphogenetic aspects of this phenomenon is very great and has been summarized by Stephenson (1930).

Despite the volume of work published on the morphological changes in regeneration, however, the amount of work available on physiological aspects of this problem is comparatively small. As late as 1948 Scheer wrote a chapter on annelid physiology without mentioning regeneration, and neither Carter (1940) nor Prosser *et al.* (1950) have much to say on this aspect of the subject.

Oligochaetes are able to regenerate both the anterior and posterior ends with almost equal facility, the part which is reconstituted depending upon where the cut is made. A cicatrice forms to seal the wound and then the remoulding of the stump proceeds, followed by the formation and growth of new segments (Stephenson, 1930) (Fig. 36). The actual process differs from species to species and the ability to provide new segments to replace missing ones is lost in some species but all individuals retain a certain degree of plasticity in the tissue.

Early morphological studies revealed that the presence of the nervous system is required for regeneration to proceed. Thus if the anterior end was removed, and the ventral nerve cord excised from the first few segments behind the injury, it was found that these latter segments are re-absorbed and regeneration takes place only from the first segments containing a portion of the ventral nerve cord (Carter, 1940). This finding may now be explicable in terms of the phenomenon of neurosecretion, and will be dealt with later.

The number of segments that can be re-formed after loss is

THE PHYSIOLOGY OF REGENERATION 119

limited, and is usually less than the number removed, but the processes governing the length and number of segments of the regenerant portion are as yet uncertain, though two hypotheses may be mentioned.

Morgan and Dimon (1904) investigated the electrical properties of the earthworm body, observing that the mid-region of the body

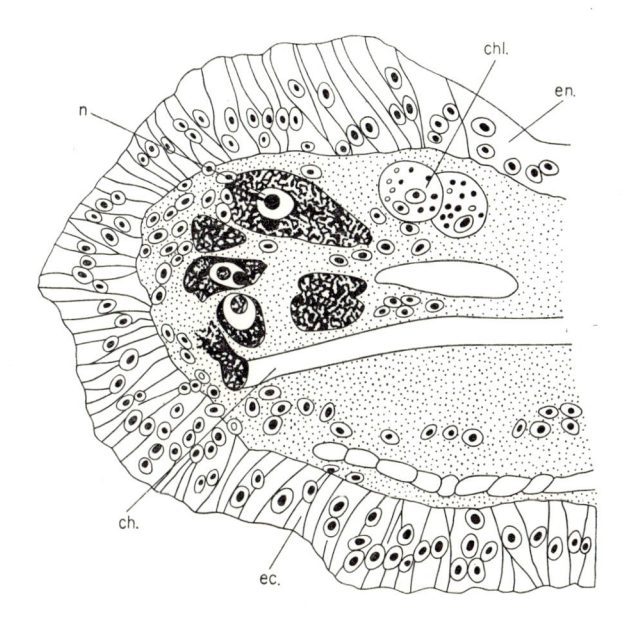

FIG. 36. *Lumbriculus variegatus:* sagittal section of a regenerating tail (1 day old), the portion ventral to alimentary canal. ec. epidermis; en. endoderm of lower wall of hind end of alimentary canal; ch, giant fibre cut in section; chl, chloragogen cell; n, neoblasts (from Stephenson, 1930).

was more electronegative than the rest of the body. They also found that when the animal was divided the cut surface was negative with respect to the adjacent tissue. Watanabe (1927) also found that the electrical potential of the body of *Perichaeta* varied from place to place, being greatest anteriorly, least in the mid-body region, and rising again towards the rear end. Similar observations have been described by Moment (1949) and Kurtz and Schrank (1955). The latter authors find the region of greatest electrical negativity to be the heart–clitellum region, which is $-16\cdot0$ mV relative to the posterior end, the mean voltage between anterior and posterior ends being in the region of $-14\cdot0$ mV, regardless of the length of the worm (*E. foetida*) (Kurtz and Schrank, see Fig. 7).

When the animal is bisected, at any level from segments 40 to 80, the electrical potential decreases sharply but returns to the original value within 3 weeks. During this period segments are replaced by outgrowth from the anterior end, but after 3 weeks regeneration growth ceases and the voltage remains steady at about the initial level. Moment (1949) therefore concluded that a certain voltage is inhibitory to the growth of further segments and this controls the length of the regenerant, a conclusion also reached by Kurtz and Schrank (1955) who noted a swing to electropositivity in the anterior–posterior gradient at the time when new segment formation ceases. Starvation for 4 days of intact animals brings about a decrease in the voltage of the posterior and clitellum regions. The anterior segments, however, are unaffected by 4 days treatment, but, in common with the other regions, declines slowly if starvation is prolonged up to 30 days. Starvation also adversely affects the rate of regeneration in *E. foetida* although the final number of segments produced is not altered since the final potential is similar in both cases.

Unfortunately there is no suggestion at present as to just how such a voltage can affect the growth and differentiation of new body tissue. Electrical stability must be governed by some process as yet unknown.

As noted previously, however, the nervous system controls the course of regeneration. If the ventral nerve cord is removed then regeneration does not occur (Zhinkin, 1936). With this in mind Massaro and Schrank (1959) have investigated the effects of nerve depressant and excitant substances upon the course of regeneration. Lithium, well known for its inhibitory effects on the growth of other organisms both vertebrate and invertebrate, depresses the number of segments that develop after injury. This action would seem to be due to replacement of sodium ions with lithium, and if NaCl is provided at the same time as lithium (3:1) then regeneration proceeds normally. Acetylcholine and the anticholinesterases, parathion and DI syston (O,O-diethyl-5-2(ethylthio)-ethylphosphorodithioate) also inhibit regeneration. These observations all indicate the importance of the nervous system in the growth of new segments, but adrenaline, 5-hydroxytryptamine, atropine, procaine and other substances that also act on the nervous system do not affect the regeneration pattern. It is thus possible that a system linked to the production of acetylcholine may govern regeneration (Massaro and Schrank, 1959). For a further discussion of the part played by the nervous system in regeneration see section on neurosecretion.

*Metabolism in Regeneration*

The metabolic processes involved in the repair of the wound caused by removal of part of the body have been traced in part.

Liebmann (1942 a, b) postulated that the role of the chloragogen tissue of the coelom was to act as a mobile repair store in case of injury. He showed that eleocytes, chloragogen cells with large basophilic granules, move into a regenerating region of *E. foetida* from undamaged segments anterior to the wound. At the site of the wound they disintegrate and discharge lipid substances into the region. This process is maximal after 9 hours and then gradually returns to normal during the next 2–3 days. The last trace of the chloragocytes in the regenerant is as a mass of phospholipid granules. The polysaccharide glycogen is also found in large quantities in the chloragocytes and may act as an energy supply to the growing area. Van Gansen (1958) has confirmed the presence of glycogen in the chloragogen cells and this evidence tends to confirm the bi-functional properties of chloragogen tissue in excretion and nutrient storage and mobilization.

Various metabolic processes in the regeneration of earthworms have been studied by O'Brien (1947, 1957b). Muscle tissue obtained from intact *Allolobophora* respired at an average rate of 14·0 $\mu$l $O_2$/hr/100 mg, varying from 11·4 $\mu$l $O_2$/hr/100 mg at the anterior end through 12·7 $\mu$l, 11·9 $\mu$l in the mid-regions to 17·2 $\mu$l $O_2$/hr/100 mg at the rear end (last ten segments). Immediately after cutting the rear end away the rate of respiration declines to about 7·0 $\mu$l $O_2$/hr/100 mg. The segments adjacent to the wound continue to exhibit a low oxygen uptake between 7·1 and 8·5 $\mu$l $O_2$/hr/

100 mg, but the regenerating tissue rises from 7·0 to 11·4 $\mu$l $O_2$/hr/ 100 mg wet weight after one week (O'Brien, 1947).

The average endogenous respiration rate of minced muscle tissue from *E. foetida* is about 30 $\mu$l. $O_2$/100 mg wet wt./hr, considerably higher than that of *Allolobophora*. When *E. foetida* was sectioned at segments 60/61 the respiratory rate fell sharply to about 10 $\mu$l $O_2$/100 mg wet wt./hr, but this was followed by a rise to a maximum of about 40 $\mu$l $O_2$/100 mg wet wt./hr, after approximately 3 weeks. From this time on values declined slowly towards normal at 10 weeks. The high respiration rates are correlated with a rapid growth rate, the regenerants adding thirty-three segments in the first 4 weeks, but only a further seven segments thereafter (O'Brien, 1957b). Enzyme activity in the form of succinoxidase does not show an early decrease but otherwise parallels the respiratory changes. During the early fast-respiring period of regenerate growth, particularly days 8–12, the blastema increases rapidly in size and differentiation of tissues goes on. It is a time when considerable increase in nucleic acids, especially RNA, occurs and it is found that barbituric acid effectively inhibits regeneration. This substance is thought to suppress nucleic acid synthesis (Massaro and Schrank, 1959).

Glycogen is present as 5·0 mg/g body weight in normally feeding animals, declining slowly upon starvation to about 3·0 mg/g body weight after 6 weeks. When *E. foetida* is bisected the glycogen content of the area very quickly falls to 2·1 mg/g in the five segments next to the wound. The preceding ten segments also have glycogen content below the normal figure. When the first small regenerant appears it has a glycogen content of only 0·2 mg/g wet wt. and it remains at a low level until regeneration is complete after 10 weeks. This suggests that glycogen is metabolized as fast as it is synthesized, preventing the build-up of stored glycogen until the growing phase is over. Glycolysis in the regenerant (10 days) proceeds at a rate 80% higher than that of normal uninjured tissues (O'Brien, 1957b) and the inhibition of glyceraldehyde dehydrogenase by iodoacetic acid (IAA) interferes with glycolysis and depresses regenerant growth. Little effect is noted initially, but inhibition of segments gradually increases up to 24 days treatment after injury (Massaro and Schrank, 1959).

This gradually increasing effect of IAA is paralleled by the maintenance of high rates of glycolysis for up to 6 weeks after section of the body, but these rates fall after that to normal at 10 weeks. The five segments in front of the cuts also show an elevation of 68% in glycolytic rates.

Associated with the accelerated process of glycolysis both lactic acid and pyruvic acid are produced, but only lactic acid increases in concentration, from 10 $\mu$g/100 mg fresh wt. in normal animals to 63·5 $\mu$g/100 mg after an anaerobic period in nitrogen. Pyruvic acid falls from 1·7 $\mu$g/100 mg to 1·0 $\mu$g/100 mg. In regenerant tissue the amount of lactic acid is double the normal after 10 days when tissue minces are incubated using glucose as a substrate.

O'Brien (1957b) thus believes that aerobic metabolism proceeds through a series of cytochrome-linked enzyme systems in a way similar to that of vertebrates. The initial drop in respiration and aerobic metabolism is due to the sudden loss of enzyme systems and their replacement by the anaerobic glycolytic mechanism, though no decrease was noted in succinoxidase activity. As glycolysis declines with the metabolism of glycogen reserves, obtained from body and chloragogen tissue, the aerobic method again takes over, rises to a compensatory peak and later returns to normal.

Other enzyme changes are described by Powell (1951). Alkaline phosphatase is often found in rapidly metabolizing tissues, and has been described as occurring in the calciferous glands of *L. terrestris*. It is also normally present in the cytoplasm of the gut epithelium of segments 35–50 but in segments posterior to 55 the enzyme is present only in the cell nuclei. Powell (1951) bisected *E. foetida* in the region between segments 50 and 51 and found that the ATP'ase contents of the five segments immediately anterior to the cuts is greatly reduced. The regenerating segments contain no ATP'ase in the gut.

This enzyme is also normally found in the ectoderm, and in the tubules and bladder of the nephridia. In regenerating segments the nephridial primordium nuclei show ATP'ase activity which spreads throughout the whole structure as it differentiates into tubules (Powell, 1951).

The relationship between morphogenetic changes during regeneration and the metabolic processes involved in tissue reconstitution have received attention from Anderson (see Collier, 1947; Anderson, 1956). *Tubifex tubifex* was found to respire at a rate of 0·16 ml/g wet wt./hr and a considerable proportion of this oxygen uptake was unaffected by potassium cyanide even in lethal concentrations. The posterior two-fifths of the body was then removed and it was found that the oxygen uptake rate remained virtually normal for the first week of regeneration. Later, however, an increase of up to 85% was noted in respiration rates, and cyanide had little effect upon any of this uptake. If regeneration was inhibited by irradiating the worms with X-rays no increase in oxygen uptake was observed. In the first week of regeneration when no respiratory increase is noted, mobilization and proliferation of neoblasts occurs, and respiration does not increase even when new segments increase in size since this happens after the late rise in respiration. It is thought that the increase in respiration reflects morphogenetic changes as it happens after initial mobilization of repair cells, but before any increase in size. As the loss of weight shown by regenerating worms is twice as great as that by intact worms kept under similar conditions it is thought that the use of body stores and metabolic cost is proportionately greater than that required to maintain the animal in a *status quo*.

The rate at which new segments are added during regeneration is affected by cyanide, $10^{-5}$M KCN initially raising the rate, $10^{-4}$M KCN depressing it. The eventual cessation of growth is not affected and the remoulded body is the same with or without KCN treatment. Worms kept in a low oxygen tension (less than 4% oxygen) in water regenerated slowly and in complete lack of oxygen no differentiation occurred at all. High oxygen tensions also blocked morphogenesis and were eventually lethal, but both inhibition and lethal effect were alleviated by simultaneous treatment with cyanide. This corresponds well with results obtained by Fox and Taylor (1955) as reported elsewhere (p. 93). Iodoacetate, an inhibitor of glycolysis, has no effect upon the rate of oxygen consumption, but increases the rate at which the new segments develop setae, so called "later differentiation". It seems, therefore, that each phase of regeneration has slightly different processes involved, and that these differ somewhat from the normal maintenance reactions of the body. In other words the demands imposed by regeneration are added onto the already present basal metabolism and require the use of specific enzyme systems to provide the energy required for morphogenesis.

## Nitrogenous Excretion and Regeneration

The processes of regeneration must obviously involve great changes in the protein metabolism of the body since these compounds provide the structural framework on which the other cellular constituents can be arranged. No work has yet been published dealing specifically with the protein picture, but Needham (1958) has presented results on the nitrogen excretion

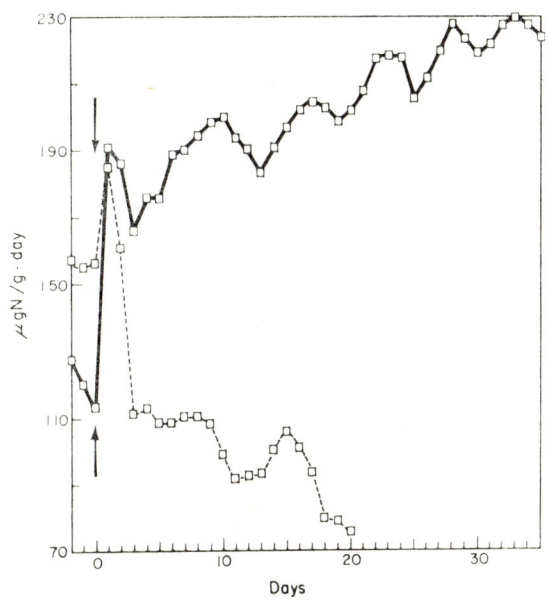

FIG. 37. Daily output of ammonia plus amino and urea nitrogen in $\mu$g N/g wet wt., smoothed by 3-day running averages □—□ by anterior halves and □...□ by posterior halves of fasting *Lumbricus* prior to and following bisection of the body (from Needham, 1958) (by permission of the Wistar Institute).

pattern of three species of earthworms undergoing regeneration of various fractions of the hind end of the body. The pattern of nitrogenous excretion obviously reflects the protein metabolism of the body.

Needham finds that the normal basal rate of nitrogen excretion is 142·1 $\mu$g/g/day for *L. terrestris*, 400 $\mu$g/g/day in *E. foetida* and 59·9 $\mu$g/g/day in young *A. longa* in fasting worms. These figures rise after removal of the posterior fifth of the body, by 19% for *L. terrestris*, 37% for *E. foetida* and 64% for *A. longa* in the first day. This increase then gradually disappears and values return to normal. The same operation on feeding animals leads to a small transient decrease in total nitrogen excretion, but the absolute values are greater. This represents a slight shock reaction which quickly passes. Isolated posterior pieces show a nitrogen excretion rate that increases over the 15 days following excision.

The increase in nitrogen excretion is proportional to the amount removed by operation and Fig. 37 shows that if the body is bisected the anterior portion shows a rapid rise in total N output, followed by a more gradual rise over 30 days. The posterior portion increases its nitrogen excretion more rapidly followed by an equally rapid fall and thenceforward shows a gradual decline till death at 20 days. The ratio of ammonia and amine N to urea N is high as in normal feeding worms (*Lumbricus*).

By removing different fractions of the same earthworm body Needham finds that there is a gradient of N excretion along the body with the greatest nitrogen output occurring towards the rear end, in the penultimate one-fifth of the body. This corresponds with several other physiological gradients reported previously in Chapter IV. The posterior peak tends to be rather further forward in fasting than in feeding worms. The details differ slightly from species to species, but the broad outlines of the situation are similar for *L. terrestris*, *A. longa* and *E. foetida*.

*Summary*

Regeneration is a common phenomenon in oligochaetes, though in some cases lost segments are not replaced. The presence of the ventral nerve cord is essential for regeneration to occur, its removal preventing regeneration of the area affected. The electrical axial field is disturbed by section but eventually recovers to the initial situation. The acetylcholine production of the earthworm body may be linked with regenerative ability. Mass migration of the chloragogen can occur after bisection of the animal. The cells break down in the area of the wound, releasing their contents. Respiration of the tissues declines upon cutting but gradually recovers. Metabolites, glycogen, lipids and pyruvic acid, for example, are affected by section. Alkaline phosphatase appears to play a great part in mobilizing energy sources to provide material for regenerating tissue. Turnover of nitrogen constituents rises greatly after cutting, the normal figure being again attained some time later. The greatest nitrogen excretion occurs at the posterior of the body.

CHAPTER IX

# NEUROSECRETION

ONE OF the most interesting and important developments in physiological work of the last 30 years has been the recognition that certain nerve cells are capable of elaborating and releasing complex organic substances acting as hormones. These cells have become modified to this purpose often to the subordination of their function as neurones. The transmitter substances released at synapses are in general localized in vesicles in these situations and act upon the immediately adjacent nerve and muscle cells. But in the neurosecretory cells the materials are often released directly into the blood system to be transported round the body and act upon some distant organ. Others act directly upon near-by glandular structures inducing them to secrete their own hormones.

Many of these neurohormone systems are now known, for example the hypothalamus–pituitary axis of vertebrates; and the sinus gland–X gland complex of crustaceans. The effects of the hormones released from these glands have actions upon chromatophores, moulting in insects, migration of eye pigments in crustacean eyes, control of heart beat (pericardial glands in crustacea), sexual cycles, and regeneration (polychaetes).

Pioneer workers in this field were the Scharrers (B. and E.). In 1937 they wrote a review of the field as it was known at that time, and already neurosecretory granules had been discovered in the cerebral ganglia of *L. terrestris*. This conclusion was based upon the observation of granules staining with particular reagents and forming vesicles and vacuoles within the nerve cells. However, no further details were forthcoming upon the number and distribution of these cells or whether the contents varied from time to time or under certain known experimental conditions.

The existence of neurosecretory cells in *L. terrestris* was confirmed by Schmid (1947) who obtained evidence for the occurrence of a cyclic variation in the activity of these cells. The application of adrenaline to the animals is sufficient to initiate the secretory cycle, and also leads to an increase in the number of cells active, but the cycle does not go to completion and only very small inclusions are to be found. Similarly novocaine increases the number of cells actively secreting.

Harms (1948) finds that in the cerebral, sub-oesophageal and the first two ventral ganglia there are two types of secretory cell: $a$-cells and blue cells. Herlant-Meewis (1956, 1957) Hubl (1953, 1956), Brandenberg (1956), Aros and Vigh (1959) and Marapao (1959) have all found neurosecretory cells in various species of earthworms. The Naididae, Lumbriculididae and Enchytraeidae all possess $a$ and blue cells (Harms, 1948).

Herlant-Meewis (1956) describes two types of cells in the cerebral ganglion of $E.$ $foetida$ which she labels $a$-cells, and $b$-cells, and associated with these is a third variety of "large and medium neurones". The $a$-cells are found in the upper posterior regions of the cerebral ganglion, and the $b$-cells, fusiform in shape, and present already when the animal hatches from the cocoon, are found at the junction of the cerebral ganglion with the circum-oesophageal commissures (Figs. 38, 39). In the sub-oesophageal ganglion lies a further type of cell ($u$-cells). Hubl (1953) gives a similar description for $L.$ $terrestris$, $L.$ $rubellus$, $A.$ $longa$ and $E.$ $foetida$, considering the "large and medium neurones" of Herlant-Meewis to be composed of two distinct cell populations. Later (1956) Hubl noted that the ventral nervous system has its own complement of neurosecretory cells; within the sub-oesophageal ganglion are the $u$-cells of Herlant-Meewis (1956), and another type he calls the $c$-cell which are also represented in other ventral ganglia. Aros and Vigh (1958) also note that neurosecretory cells are present in the sub-oesophageal ganglion and the other ventral ganglia, and that these latter are organized into groups lying symmetrically on the right and left sides of the nerve cord.

Hubl (1956) believes that the $c$-cells may represent $b$-cells in a different stage, and Marapao (1959) describes three secretory stages in one cell and further shows that the two cerebral types are independent of one another.

Production of neurosecretory hormones has often been associated with cyclic bodily functions such as reproduction, pigment

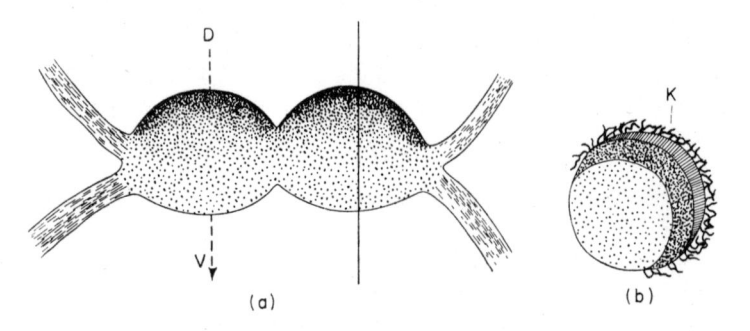

FIG. 38. (a) Cerebral ganglia of earthworm; D = dorsal; V = ventral; (b) section through ganglion showing region of ganglion cells, hatching indicates area containing *a*-cells and capillary network (K) (from Hubl, 1953).

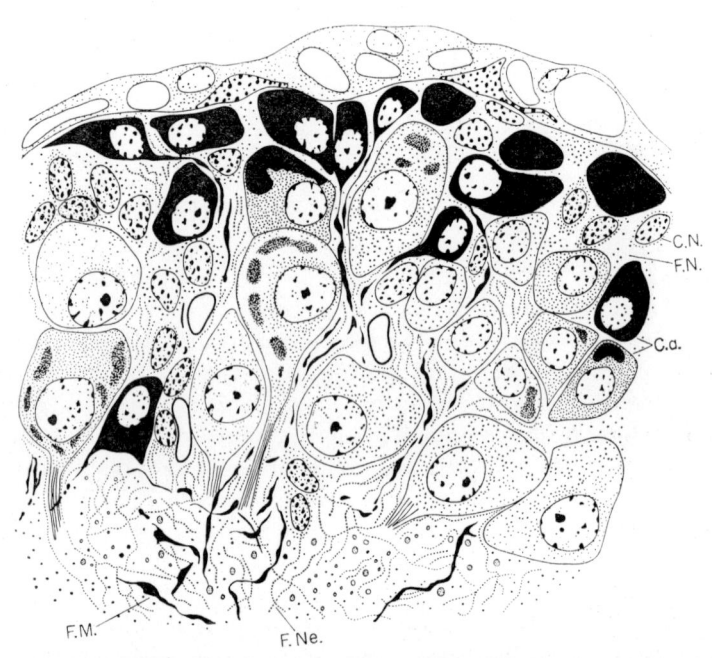

FIG. 39. Section through posterior region of cerebral ganglion of *Eisenia foetida*. C.a. = *a*-cells forming neurosecretion; C.N. = neuroglial cells; F.N. = neuroglial fibres; F.M. = moniliform fibres; F.Ne. nerve fibres (from Herlant-Meewis, 1956).

migrations, and development of secondary sex characters. Oligochaetes exhibit activity cycles, diurnal respiratory cycles, annual reproductive cycles and in some cases a periodic aestivation. Some or all of these functions may be controlled by neurosecretory methods though evidence is still wanting in many cases.

For example Abeloos and Avel (1928) worked upon the regeneration of the body of *Allolobophora* (two species). The anterior end of the body can be replaced at all times of the year, but the posterior extremity can be regenerated at certain periods. This latter function is an annual cycle concurrent with diapause (aestivation) and they postulated that it is probably regulated by some internal mechanism. Bailey (1930) also found that the regenerative ability of *E. foetida* exhibits an annual cycle.

Michon (1949) has since shown that aestivation is controlled to a large extent by the conditions of soil moisture and humidity. As the water available to the animals is decreased so they are more likely to go into a condition of diapause, but Michon was unable to explain why it is not possible to reawaken the animals when water is provided in abundance during the diapause period. Once the earthworms are aestivating a certain time must elapse before they again become active. This is suggestive that a hormone is involved regulating the activities of the species by an annual secretory cycle. Long before this Avel (1929) had shown that when the temperature is maintained constant, humidity is kept at a satisfactory level, and feeding is continued, earthworms will go into diapause anyway. This occurs at the end of spring when the reproductive period is over and is accompanied by a loss of sexual characteristics such as regression of the clitellum.

Even earthworms that do not undergo aestivation e.g. *L. terrestris*, exhibit a yearly reproductive cycle, the gonads coming to maturity in spring and early summer and regressing later in the year to redevelop in the following spring. Of the types of secretory cells in the cerebral ganglia one, the *a*-cell, is apparently involved in this periodic change, but the *b*-cell is not. Individuals in which the reproductive system has not yet matured are lacking in *a*-cells, but they differentiate as the animal ages. In adult animals exhibiting the yearly maturation of sex organs the *a*-cells discharge their contents in spring and the number of cells decreases. If the gonads are extirpated the colloid content of the *a*-cells decreases as the cells discharge into the capillary network surrounding the ganglion, presumably mobilizing the materials necessary to regenerate the sexual apparatus (Hubl, 1953).

More extensive experiments on this subject were carried out by Herlant-Meewis (1957) with regard to the effect of the nervous system on reproduction, egg laying and cocoon production. She surgically removed (a) the cerebral ganglia, (b) the sub-oesophageal ganglia, (c) the circum-oesophageal collar and (d) the ventral nerve cord between segments 4 and 6/7. In case (a) the worms lose weight, and the secondary sex characters such as the tubercula pubertatis disappear and egg laying ceases. Egg laying begins

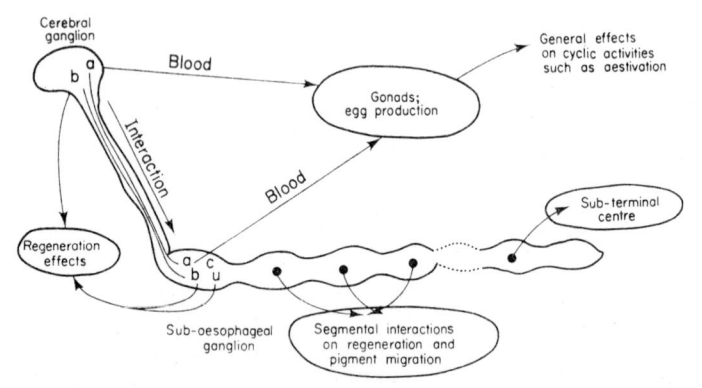

FIG. 40. Possible pathways of neurosecretory interactions in an earthworm. Letters indicate distribution of cell types.

again after a delay of 8 weeks during which time the cerebral ganglia regenerate. When the sub-oesophageal ganglia are removed weight is lost only in the second week after the operation and is regained in the fourth week. Cocoon productive capacity is lost for 2 to 15 weeks. Extirpation of the entire oral complex of the nervous system interrupts egg laying for 7 to 17 weeks, and removal of the ventral nerve cord immediately behind the oral region stops production of eggs for 3 to 8 weeks. This last period is of the same order as the delay in egg laying shown by controls in which the body wall was opened but none of the nervous system was removed.

Histological examination of the ganglia from animals at various times during these processes indicates that two separate systems are involved. The removal of the sub-oesophageal ganglion is followed by an accumulation of secretory and colloidal material within the *a*-cells of the cerebral ganglion. This leads to the suggestion that the cells, known from the work of Hubl (1953) to affect reproduction, do so not by a direct action upon the gonads, but by an indirect method, first stimulating other neurosecretory cells within the sub-oesophageal ganglion to discharge their products. Hubl (1953) states that the *a*-cells discharge into the capillary network surrounding the cerebral ganglion and as this is interrupted when the sub-oesophageal nerve mass is removed there is no continuity between the two neurosecretory systems. The secretions are evidently not released into the coelom since this is always in contact with both nerve concentrations and is not interrupted by the removal of either, or by severance of the circum-oesophageal commissure. Removal of the cerebral ganglion is likewise followed by the appearance of concentrations of secretory materials within the sub-oesophageal ganglion. This material sometimes also makes an appearance in the axons leading from this gland. Egg production in this case ceases immediately.

To summarize then it is evident that the *a*-cells of the cerebral ganglion play a regulatory part in the reproductive cycle of the earthworm. They show a regular cyclic production of secretory material within the cells and release their products into a capillary blood supply in the spring-time when the gonads come to maturity, their action being mediated through a second set of secreting cells in the sub-oesophageal ganglia. Marapao (1959) has also adduced evidence for a hormonal transmission of brain substances since injection of brain extracts mimics the effects obtained from intact animals.

Although it is not to be supposed that changing day length has any effect upon the neurosecretory activities of the ganglia of earthworms it is interesting to note that sunlight, ultra-violet illumination and darkness all affect secretory cells (Aros and Vigh, 1959) (see Törö, 1960). The medio-dorsal cells ( = *a*-cells of Hubl?), and the lateral secretory cells in the segmental ganglia show vacuolization and discharge when the animals are placed in visible or ultra-violet light. Replacement in the dark leads to a further manufacture of hormone and the gradual refilling of the cells with droplets. When the cells are evacuated the droplets are found distributed along the axons and nerve fibres. When earthworms are exposed to light, something which does not happen very often in the field, they show a progressive deepening in colour, and Aros and Vigh (1959) suggest that the neurosecretion is associated with a chromatophorotrophic response. However, no chromatophores or pigment containing cells that are able to expand and contract have as yet been described in earthworms. It is known, moreover, that the pigments of the integument contain protoporphyrin and its methyl esters. Porphyrins are photodynamic and react strongly to light, and in mammals can lead to severe necrotic conditions of the skin. It is thought (Merker and Braunig, 1926) that ultra-violet light is lethal to *E. foetida* and the animals disintegrate after exposure to this type of light; sunlight is also lethal to earthworms, probably by virtue of the ultra-violet component. More recent experiments (Satchell, personal communication), however, indicate that ultra-violet light is not a lethal factor. It is possible that the darkening of earthworms in light is due to photosensitization of protoporphyrin rather than some neuroendocrine response.

One of the most interesting phenomena encountered in annelids in general, and of which the oligochaetes are a good example, is the ability to be able to replace large portions of the body that are lost by accident or design. Both anterior and posterior regions can be regenerated, the portion being replaced depending upon the position at which transection occurs. If the whole oral complex of ganglia is removed it is only a matter of time before they again appear. But if the ventral nerve cord is removed from the segments immediately adjoining the cut, regeneration takes place only from the cut surface of the nerve cord and not from that of the cut segments (Carter, 1940). Bailey (1930) also found that if the ends of the nerve cord are reflected away from the normal direction of growth then regeneration does not start and the cut surface simply rounds up. Similarly it is possible to induce the formation of a new bud at the site where the nervous system is reflected growing laterally from the body. It is evident, therefore, that the nervous system plays a vital role in the growth of new tissue.

There is evidence from the amphibia that regenerative ability is a function of the nerves themselves influencing the surrounding tissues without the implication of other mechanisms. But in annelids there are indications that a neurosecretory process is involved as well.

Harms (1948) found that the cerebral ganglion has a humoral effect and that the anterior nervous system is essential to the process of regeneration, suggesting that this is due to the necessary neurosecretory cells being limited to this region.

Hubl (1956) removed the cerebral ganglion and the posterior extremity of the animal at the same moment, observing that the rear end will not grow again, thus confirming Harms's suggestion that the anterior nervous system is essential for regeneration. This action is an inhibitory one. Regeneration can occur if a short period elapses between the operations. In 1953 Hubl showed that the $b$-cells of the cerebral ganglia are involved in regeneration and in 1956 he considers that they are inhibitory to the process of regeneration, as also are cells in the central segmental ganglia. When $b$-cells are removed a short time after the hind end is removed the inhibition is released and growth can take place. Excision of parts of the nervous system further to the rear e.g. the sub-oesophageal ganglion and the adjoining ventral nerve cord, also inhibits new growth, whilst loss of the ventral nerve chain from body segments has a lesser inhibitory effect.

The secretory cells of the nervous system have complementary effects upon one another. The cells of the sub-oesophageal ganglion affect the $b$-cells of the cerebral ganglion. The cells of the ventral nerve cord show a gradient of activity, being more productive anteriorly than posteriorly. The $u$-cells contained in the sub-oesophageal ganglion seem to be causal agents in promoting regeneration, and removal of this ganglion inhibits growth since the appropriate cells are no longer present. This is in contra-distinction to the effect of the $b$-cells which are inhibitory when present and allow growth to occur when they are removed.

In cases where the posterior end of the body is removed without disturbing the anterior end some other stimulatory mechanism must invoke activity of the $u$-cells (sub-oesophageal ganglion) in order that regeneration can occur, perhaps removal of the $c$-cells of the ventral ganglia at the rear upsets the normal hormonal balance between the various ganglia, and thus induces secretion in the anterior ganglia. As regeneration proceeds more $c$-cells are formed until once again the hormonal balance is restored and growth ceases.

These facts do not, however, explain why the nervous system must be intact to induce regeneration within a particular segment, unless secretory products tend to be released locally from circulation throughout the body. In the latter case stimulatory substances could be distributed to segments at a distance from the site of origin and thus be stimulated.

This explanation should be compared with the hypothesis of the development of a critical inhibitory voltage due to Moment (1949) and Kurtz and Schrank (1955) (see Chapter VIII).

A sub-terminal centre in the posterior region of the body has been suggested by Needham (1958) on the basis of studies in the nitrogen excretory pattern during regeneration. But apart from the segmentally repeated neurosecretory cells of the ventral nerve cord (Aros and Vigh, 1959) there is no histological or cytological evidence at the moment for this.

There is no information at the moment as to the chemical nature of the products of these secreting cells. Pharmacological studies point to the implication of adrenaline, acetylcholine, and possibly 5-hydroxytryptamine as transmitter or more widely acting chemicals deriving from the nervous system (see Chapter XI), but nothing has so far implicated these substances in the sexual or regenerative growth processes. Schneidermann and Gilbert (1958) found that extracts from *L. terrestris* contain substances that have insect juvenile hormone activity, maintaining pupal skeleton in a treated area when *Anthereae polyphemus* metamorphoses to the adult moth. But the structure of this active substance is unknown, and there is no evidence as to what, if anything, it does in the earthworm.

*Summary*

Neurosecretory cells were early discovered in *Lumbricus* and their presence confirmed many times since. Various types of neurosecretory cells have been described and some of these have been associated with physiological functions. In particular, effects of the secretory areas have been noted on activity of the reproductive organs, egg laying and cocoon production, with possible suggested effects on aestivation and other rhythmic activity. Aspects of regeneration phenomena may also be governed by the activity of neurosecretory cells lying in the anterior ganglia and those of the ventral chain. Physiologically active substances have not yet been isolated.

CHAPTER X

# NERVOUS SYSTEM

In the early years of this century experiments on the nervous system were made in many invertebrates, relying upon macroscopical studies on the effects of nerve section on behaviour, what happened when the nerves regenerated and how the behaviour altered in response. Histological investigations gave details of the innervation of sense organs and motor systems and further experiments revealed something of the properties of these systems.

Since the 1930s, however, our knowledge has increased rapidly with the introduction of electro-physiological techniques. With the advance of amplifiers, oscilloscopes, stimulators, tape recorders and all the other impedimenta of the neurophysiologist, our knowledge of nervous activity has increased rapidly. Some of the first experiments were carried out on invertebrates and the first microelectrode implanted into a single axon was placed in the squid giant axon. A large volume of work has accumulated and is excellently summarized in the recent book by Bullock and Horridge (in press). By far the greatest volume of work has been carried out on insects, crustacea and molluscs, but other phyla have not been ignored.

In the realm of earthworm physiology the functions of the nervous system have received more attention than any other aspect. Much of the interest in this system has centred upon the presence of three giant fibres which extend the entire length of the body and conduct very rapid impulses from end to end. And yet despite the considerable experimentation made upon them some of the properties of these fibres still evade description. In other aspects very little is known about sense organs, synaptic properties, motor fibres and innervation of muscles. It is not proposed here to go deeply into the anatomical and histological structure of the nervous system since very full descriptions can be had from

Smallwood (1926, 1930), Prosser (1934a), Hanstrom (1928), Hess (1925a), Langdon (1895) and Stephenson (1930), but an outline may be of use in pinpointing the structures mentioned later.

The main nerve trunk of the body runs ventrally from segment 4 to the rear end, and has one ganglion, a collection of nerve cells, to each body segment. Anteriorly the cerebral ganglion is a bilobed structure found dorsally in segment 3, giving rise to two stout nerves that pass forward to the prostomium which is well supplied with sense organs (Langdon, 1895). The nerves supplying the two most anterior segments arise from the circum-oesophageal commissures which envelop the oesophagus in segment 3. Segment 4

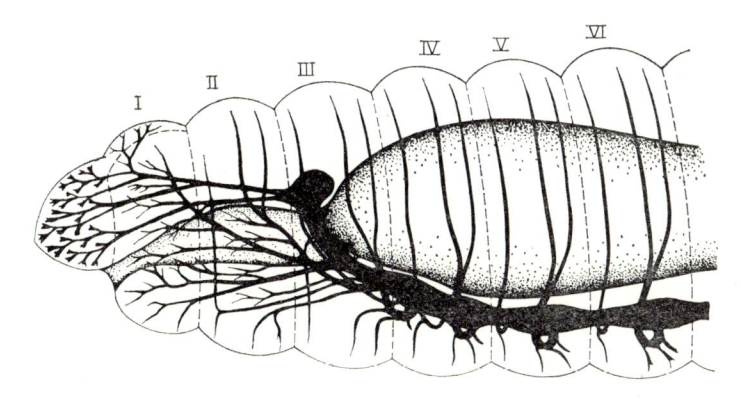

Fig. 41. Anterior nervous system of *Lumbricus terrestris*. I–VI = segments 1–6 (from Hess, 1925).

contains the sub-oesophageal ganglion which represents the fused ganglia of segments 3 and 4, and the innervation of both segments stems from this ganglion. In a typical body segment three pairs of nerve trunks originate from the ventral nerve cord and pass outwards to the body wall. The anterior pair arise just behind the anterior septum of the segment anterior to the ganglion and at a distance from the other two pairs which stem from the segment ganglion and are close together. The third nerve on each side gives off a small branch to the septum at the posterior side of the segment. In the caudal segment, which apparently is the equivalent of two body segments, five to six nerves are found (Hess, 1925a) (Fig. 41).

The gut is supplied with a plexus which lies within the intestinal wall. The plexus originates from the circum-oesophageal commissure where six small nerves leave the ring and send branches along the gut wall. The muscles of the intestinal wall are supplied by nerves arising in the septal plexus of each segment. In the body

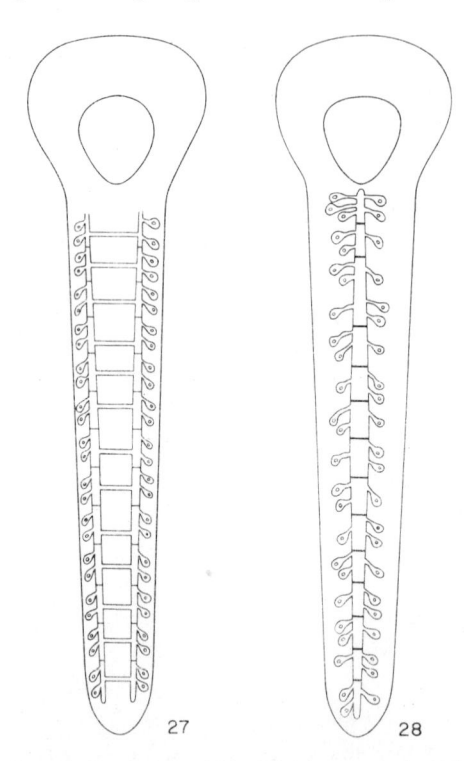

FIG. 42. Diagrams of paired lateral giant axons and the single median giant axon of *Lumbricus* (from Nicol, 1947).

wall branches of the motor nerves pass to the circular and longitudinal muscle layers, and sensory fibres run from the sense organs on the surface of the body. A sub-epidermal network of nerve fibres has also been described (Smallwood, 1926).

The relations that exist between the various fibres that comprise the nerve trunks are indicated in Fig. 43. This diagram is based upon anatomical studies, and at the present time little is known of the physiological properties of the connections existing between the various components of the reflex arc, apart from the giant fibre "startle" response. From Fig. 43, however, it can be seen that sensory axons, the cell bodies of which are in the epidermis, pass to the ventral nerve cord, there to make synaptic connection either directly or via an association neurone, with motor fibres. As a result of sensory stimulation in one segment a motor response may occur via fibres in the same segmental nerve, the contralateral nerve of the pair, the nerve before or after that stimulated, or even the segment

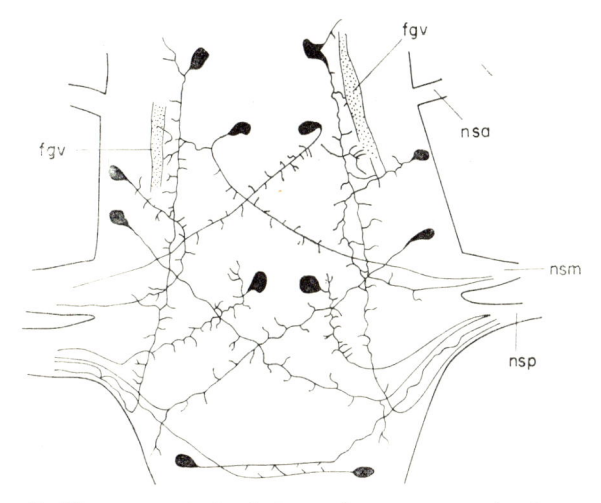

FIG. 43. The anatomical relations of motor axons in the ventral nerve cord of *Eisenia*. fgv = ventral giant fibres; nsa = anterior segmental nerve; nsm = median segmental nerve; psn = posterior segmental nerve (from Grassé, 1949).

before or behind. Connections via association neurones extend for up to three segments from the one stimulated (Stephenson, 1930, Hanstrom, 1928). Alternatively if the stimulus is strong enough the association neurones may be by-passed and the giant fibre rapid response brought into action.

Not only can responses be spread from segment to segment by neurones within the ventral nerve cord, but the sensory fields of the nerves supplying the body wall also cover more than one segment. In the course of an electrophysiological investigation Prosser

(1935) discovered that tactile stimuli, giving rise to action potentials within a segmental nerve, could be perceived as giving rise to activity when touch was applied in the segment of origin or in the segments to either side as well, though the areas on the segmental wall to fore and aft were rather more restricted (Fig. 44). Thus, at least for the touch receptors, the sensory nerves ramify over a number of segments beside that in which they originate. At the same time it is thought that, whilst the sub-epidermal nerve network is organized on a segmental basis, the various forces applied to the body wall by movement of contiguous segments is translated into nervous activity in this system, thus enabling peristalsis to continue uninterrupted from segment to segment even when the ventral nerve cord is severed.

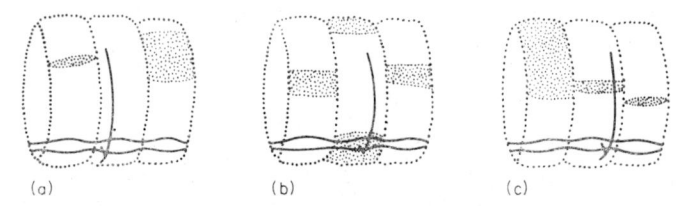

FIG. 44. Sensory fields served by (a) first, (b) second, (c) third segmental nerves. Dotted areas are those from which no responses were obtained when explored by a fine needle (from Prosser, 1935).

There are many sense organs of various types within the body wall of *Lumbricus* and these are associated with numerous intra-epidermal nerve fibres that end freely between the epidermal cells (Langdon, 1895). Langdon believed that each sense cell had its own individual nerve fibre that ran as a separate entity into the ventral nerve cord, but as there are thousands of sense organs per segment, and only a few hundred fibres in each segmental nerve (150 in *L. terrestris*, Smallwood, 1930; 1500 in *Perichaeta*, Ogawa, 1927), and in the ventral nerve cord ganglia, it is obvious that this cannot be so. Smallwood (1926) suggested that the sense organ fibres synapse with the sub-epidermal network of fibres and are collected together from there to form the segmental nerves. By the same token the motor fibres of the ventral nerve cord, relatively few in number are also subject to splitting into many fractions in order to supply each muscle fibre within the thick longitudinal and circular muscle blocks. Horridge (private communication) has,

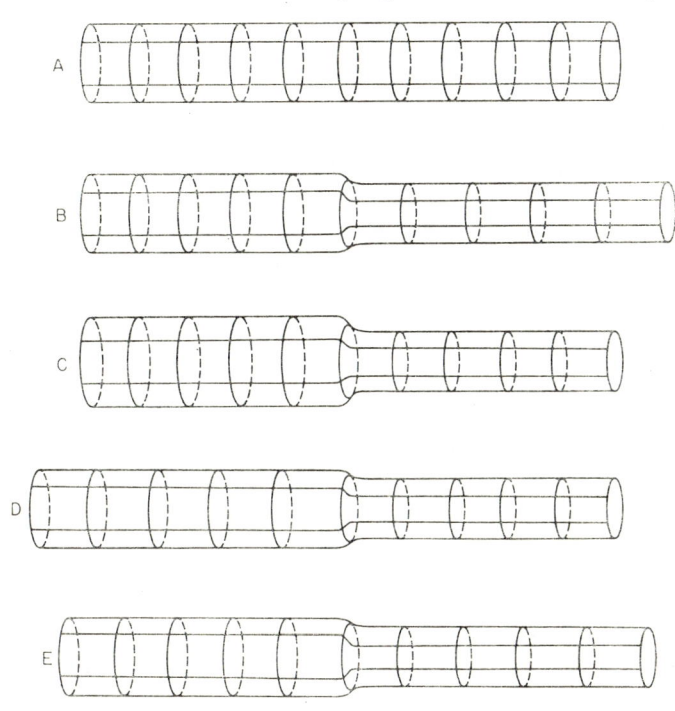

Fig. 45. Diagram to illustrate four possible results of the contraction of circular muscles at one end of a cylindrical animal. In A the muscles are all relaxed. In B the circular muscles of the right-hand end have contracted and this end has elongated. The left-hand end remains the same. In C the length of the right-hand end has remained the same but the diameter of the left-hand end has increased. In D the length of the right-hand end has also remained the same. The length of the left-hand end has increased but not its diameter. In E the length of both ends has increased but their diameters have remained the same as in B and D (from Chapman, 1958).

however, shown that in the polychaete, *Harmathöe*, the segmental nerves contain very many neurones, distinguishable only with the electron microscope, and the same state of affairs may be true of *Lumbricus*.

*Peristalsis, Co-ordination, Movement and Locomotion*

*Co-ordination of the body wall musculature.* When an earthworm crawls along the surface of the ground, or through the channel that comprises the burrow in which it lives, it does so by the co-ordinated contractions of the longitudinal and circular muscle bands that go to make up the bulk of the body wall. From a position of rest the contraction of the circular muscles may have one of four effects: the contracting end may elongate, the opposite end may elongate or it may thicken, or both ends may elongate. Which event takes place depends not on the contraction of the circular muscles of the contracting end but upon the state of contraction of the longitudinal and circular muscles in other parts of the body (Chapman, 1958). These actions are seen in Fig. 45.

Much of the muscular reactions of the body wall depends upon the presence within the segments of coelomic fluid. Each segment is in effect a self-contained sac of fluid, for although the septa between the segments are incomplete ventrally where the nerve cord passes through, it is possible to seal off this hole by a sphincter muscle around the nerve cord. Injection of dyes into individual segments shows that no transmission of fluid from segment to segment occurs when peristaltic waves pass along the body. If the animal is anaesthetized and muscle tone lost as a result then coelomic fluid may pass between segments. The average hydrostatic pressure amounts to about 16 cm water in the anterior segments, and approximately 8 cm water in the rear segments, although greater values may be reached during the rapid wriggling movements sometimes shown by worms. The pressure exerted by the earthworm prostomium upon the ground is greatly increased when the animal is in the enclosed space of its burrow (Newell, 1950). Roots and Phillips (1959) have shown photographs of the method by which the prostomium gains the necessary purchase for burrowing.

In a period of quiet locomotory activity an earthworm moves along by first contracting the circular muscles of the anterior end, causing the anterior segments to extend. The contracting region passes posteriorly, and as the pressure lessens the anterior longitudinal muscles contract, drawing the posterior end forward. The posterior region is unable to slip backwards for the chaetae of the body wall are aligned to point backwards. The first wave of contraction is followed by others in a rhythmic succession and the worm progresses slowly forward (Friedländer, 1894, Beidermann, 1904). Chaetal movements are carried out by a special musculature. A retractor muscle links the inner ends of the two chaetal pairs on each side of each segment. Protraction is carried out by a tent-like arrangement of muscles running from the inner ends of the chaetae to the body wall. There are no special muscles to rotate the chaetae on their long axes, as is necessary when direction of locomotion is reversed—this is done by differential action of the protractors, assisted by the pressure exerted on the chaetal pairs when the

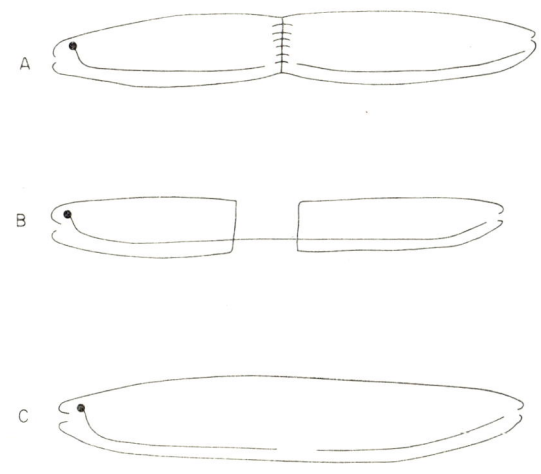

FIG. 46. Situations affected by Friedlander (1894) and Beidermann (1904). (A) animal bisected and rejoined; (B) body cut, ventral nerve cord intact; (C) nerve cord severed, body uncut. Peristalsis occurs synchronously in all three cases.

longitudinal body wall muscles contract (I. Linn, personal communication).

Three early experiments indicated that the propagation of this peristaltic wave from segment to segment is due to a combination of factors. If an earthworm is completely severed and then the two fragments rejoined by sewing thread, a peristaltic wave starting at the anterior end progresses back without interruption at the cut surfaces (Fig. 46). If now another animal is cut so that only the ventral nerve cord remains intact the waves of the contraction still pass from anterior to posterior without hindrance, the impulses necessary for this action passing along the ventral nerve cord for at least thirty segments at a rate of about 25 mm/sec. Thirdly, if the ventral nerve cord is removed completely from several segments leaving the rest of the segments intact, the co-ordination of front and rear is not affected (Friedländer, 1894; Beidermann, 1904; Garrey and Moore, 1915; Moore, 1923a). It is therefore apparent that the peristaltic wave can be propagated from segment to segment either along the nerve cord via the intermediary neurones, or by mechanical stimulation of the succeeding segments.

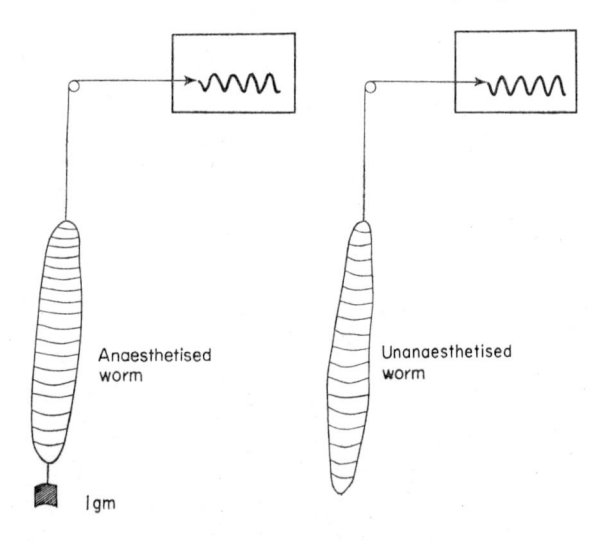

FIG. 47. Effect of weight on muscular movements of *Lumbricus* (after Moore, 1923a).

Gentle tactile stimulation of the surface of a worm will induce movement of the body, but anaesthesia with $MgCl_2$ blocks this response. If such an anaesthetized worm, or piece of worm, is now suspended vertically and weight is applied to the posterior end a peristaltic wave is set up (Fig. 47). Thus both touch and tension induce contraction though tactile stimulation first causes extension, whilst tension causes contraction. If the experiment is so arranged that the ventral nerve cord alone is stretched no peristalsis begins. This leads one to suspect the presence of intradermal receptors, possibly in the deep muscle layers. Dawson (1920) has described four types of nerve cell lying in the muscle layers. Two of these types may be of interest here. The first of these is a large cell lying interposed between the ventral and dorsal chaetae of each side in each segment, and apparently ideally placed as tension receptors since axons leave these cells and run towards the chaetae. Dawson, however, was unable to find evidence for axons running centrally to the nerve cord and therefore believed these cells to be concerned with effector reactions. The second type of cell lies in the circular muscle layer, and is similar in many ways to sensory cells found on the surface of the body. "Just what function these deep-lying cells perform is difficult to surmise. Their position in the circular muscle layer and their possible restriction to the ventral region indicate a probable role in connection with the initiation or maintenance of the creeping movements of the worm", Dawson (1920). This percipient viewpoint was expressed long before the physiological demonstration of such cells as stretch receptors which are now known to be widespread among arthropod and vertebrate muscles. It would be interesting to examine the electrical properties of these earthworm cells by electrophysiological means, though a rhythmically discharging cell that may be of similar type has been discussed by Laverack (1960b), and Prosser (1935) found that pulling the body wall gave rise to action potentials in the segmental nerves. The assumption of a tension or stretch receptor function for these cells provides an adequate basis for an explanation of the phenomena of co-ordinated movement occurring in animals in which the continuity of the nerve cord has been interrupted. The normal course of impulses runs through the nerve cord and via the segmental nerves to the body wall, but if the CNS is severed the passive tensile and tactile stimulation caused by dragging the rear segments across the rough substratum will be sufficient to start up fresh reflex activity in these segments. It should be noted moreover that the normal pattern of movement is obtained only when the body touches the ground, and not when it is suspended (Moore, 1923a).

It is possible to demonstrate that the tension reflex of pieces of the body wall have a distinct threshold which ranges from $0.1$ to $1.0$ g. That is to say the smallest weight needed to evoke regular peristaltic movements is of this order, though it naturally varies from preparation to preparation and from time to time. This can be demonstrated when the earthworm fragment is hung from an isotonic lever, the animal being immersed in saline to counteract the weight of its own body.

It is essential that the weight should be applied for a certain length of time, the threshold of duration. If weights are removed before this period has elapsed then no peristalsis is set up, but once contractions start they may continue for as long as 3 minutes in response to one stimulus. This sort of response is typical of reflex action and is known as "after-discharge". This is brought on by a persistence of activity within the reflex centre even after the external stimulant is removed. This is demonstrated by the

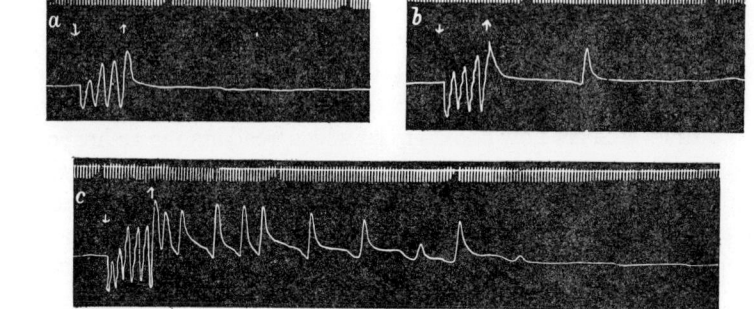

Fig. 48. After discharge invoked by the application of various weights for the same length of time (15 sec). (a) 0·5 g; (b) 1·0 g; (c) 1·5 g. Weight applied between arrows (from Collier, 1939).

observation that the after-discharge period is directly proportional to the magnitude of tension applied if this is done for a constant short period of time (Fig. 48). There is now some neurophysiological evidence for this belief in the earthworm (Kao, 1956) which will be discussed later.

When the stretch stimulus is maintained continuously for long periods the response in the reflex arc falls off after about 25 minutes. This is probably due to adaptation, and consequent raising of the threshold, of the sense organs to a given set of conditions since applications of a greater weight, and hence a greater tension, will renew peristalsis (Collier, 1939 a, b). Though the tension reflex may be of great importance in maintaining a peristalsis throughout the body it cannot initiate such a contraction. Peristalsis starts at the anterior end of the body and passes backwards. Any stretch receptors present in the anterior segments cannot be stimulated when the animal is at rest and it needs muscular contraction to occur before these organs can respond. Such muscular movements happen only when the animal is already moving, so the stretch receptors can only prolong a movement already started by some other agency.

Removal of tension from the sense organs is followed by a cessation of peristalsis. If a worm is allowed to hang by its own weight and contracts rhythmically, the rhythm can be stopped by surrounding the animal with water (Gray and Lissmann, 1938).

The tension receptors may also be concerned in the homostrophic response. An earthworm is placed so that the anterior end is upon a stationary surface, but the rear end rests on a moveable slate. If the slate is deflected to one side the front end of the worm turns until it is parallel to the direction of the rear end. The number of segments involved in the response of the anterior end depends upon the number of segments displaced at the hind end up to a limit of about twenty segments. The receptors for this response are present along the entire length of the body and transmission of the impulses takes place via the ventral nerve cord, transection of which prevents the response occurring. The effectors for the homostrophic response, however, are confined to the anterior fifteen to twenty segments. It is suggested that this reflex is essential in enabling the earthworm to maintain a stable position in space. No statocysts or other type of gravity receptor have been described in earthworms so the activity of stretch organs and their effectors seems of great importance in informing an earthworm whether it is the right way up or not (Moore, 1923a), although it may not explain how *Lumbricus* is aware that it is proceeding head uppermost to feed, or *A. longa* tail uppermost when casting.

The after-discharge process noted in the tension reflex is sustained, as mentioned above, for some while and there is some evidence relating to this, suggesting that it may be due to the action of transmitter substances lingering in the system and causing prolonged firing of motor nerves. Kao (1956) has found evidence within the giant fibres for prolonged sensory bombardment and electrical inputs continuing for some while (see p. 163).

Pharmacological evidence as to the response of the body wall to various transmitter substances shows that there is a relative lack of sensitivity to acetylcholine, the threshold being between $10^{-4}$ g/l., compared with the activity of the gut. The body muscle response is greatly potentiated by eserine which lowers the threshold to $10^{-6}$–$10^{-8}$ g/l. As eserine inhibits the action of cholinesterase by competitive action this prompts one to postulate a fairly large concentration of this enzyme present in the normal earthworm body wall preventing acetylcholine action (and also in the body wall of *Hirudo* and *Arenicola*). Nicotine causes a single large contraction and completely abolishes the muscular response to acetylcholine in high concentrations. This, coupled with the observations of Mendes and Nonato (1957) that acetylcholine causes vasoconstriction in the body wall suggests that acetylcholine may be a transmitter in the peripheral nervous system.

*Electrophysiology*

The electrical activity of the nervous system associated with the normal locomotion of earthworms has only once been described, by Gray and Lissmann (1938). Recordings were made from the ventral nerve cord along the length of which, adjacent to the recording site, the segmental nerves had been cut. When the animal was still no nerve action potentials were seen in the central nervous system. If peristalsis started, however, a well-defined rhythm of impulses was recorded, the frequency of which was identical to that of the body movement. So far as could be judged the electrical potentials coincided with the phase of longitudinal contraction but the observation is not conclusive. It is less easy to explain the spontaneous activity of the nerve cord when isolated from the body. Rhythmic bursts of potentials occur as in the *in situ* recordings, but in the majority of cases the frequency and duration of the bursts is not like that of peristalsis. The variation in rhythm may be partly due to independent movement of the isolated cord on the electrodes, since the nerve fibres are enclosed in a muscle sheath which contracts with a rhythm of its own.

Although these electrical rhythms seem to be the basal nervous activity of the animal Gray and Lissmann (1938) were unable to conclude that these initiate movement in the earthworm. They offered two explanations, between which they could not choose.

First that longitudinal contraction is initiated by the tactile sense organs on the ventral surface of the body and which functions only when in contact with the substratum, or secondly, that transition from circular to longitudinal contraction is affected by a central nervous mechanism which is inactive in the absence of an adequate level of peripheral excitation, as from stretch receptors. In the light of their neurophysiological findings the problem was unsettled, and remains so at present. (Once movement has started methods of passing the stimulation from segment to segment are as explained above, but how these movements begin has not yet been elucidated.)

*Properties of the Reflex Arc*

The properties of the synaptic junctions lying on the reflex arc, (body wall, sense organs, ventral nerve cord, body wall muscles), have recently been outlined by Roberts (1960). Using electrical methods of stimulation he examined the synapse between the sensory and the giant fibres, between the giant and the motor fibres, and between the motor and muscle fibres (the neuromuscular junction).

By repetitive stimulation of the central end of a segmental nerve he recorded potentials from the muscles of the stimulated segment. A potential was obtained after the first shock, growing in size during the first few shocks in the frequency range of 1–10/sec. After this the size of the potential remains constant for about eighty shocks, and then gradually decays until after 700 shocks it is but a third of its original size. If stimulation is continued even longer the potential may disappear altogether, but the synapse recovers quickly and within 2 minutes the muscle potential is full size again, though fatigue is more rapid when the stimulation is repeated over a long perid. The neuromuscular junction then is capable of fairly long periods of firing without interference, but eventually fatigues. The procedure, however, is not short enough to account for the rapid exhaustion of the shortening "startle" reflex of the earthworm which is co-ordinated by the giant fibres (Kuenzer, 1958; Roberts, 1960).

The central synapse between the giant fibre and the motor nerve can be studied by stimulating the giant fibre anteriorly, and monitoring the impulses from muscles midway along the body. Roberts (1960) found this a difficult operation though for the synapse is apparently very easily fatigued, being stimulated to failure during the dissection of the preparation. Even in successful experiments the accommodation of the giant fibre: motor fibre junction was complete as soon as the giant fibres were stimulated. But in some cases muscle action potentials were seen after a 25 msec delay. After passage of only a few shocks (at 1/sec) in a 1:1 ratio the transmission across the synapse became irregular for 5–20 sec and then ceased to function (Fig. 49). The size of the

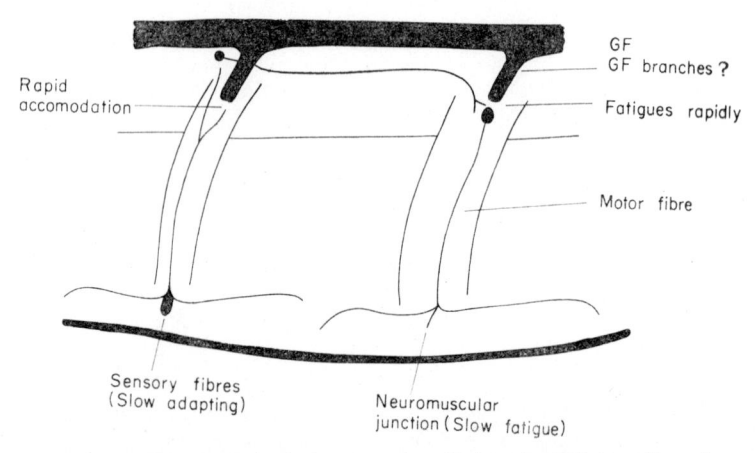

FIG. 49. The physiological properties of the giant fibre reflex of *Lumbricus* (after Horridge and Roberts, 1960 and Roberts, 1960).

potential recorded varies during this period, presumably indicating that more than one motor fibre is present in each segmental nerve. In response to the same electrical stimulation a wave of slower, smaller action potentials can be seen, and the accommodation of the junction to these impulses is less rapid. The conduction velocity of the motor fibres in the segmental nerves is of the order of 0·4 m/sec, suggesting that the giant fibre or some other rapidly conducting system does not operate in the segmental nerves, unlike the situation in some polychaetes such as *Myxicola* (Horridge and Roberts, 1960). Recordings of the size of potential obtained from the segmental nerve in relation to giant fibre spikes, however, seems to show that the giant fibres may extend a short way into the central end of the segmental nerves and there make synaptic contact with other motor fibres (Laverack, unpublished observations). Histological studies by Smallwood (1926) and Stough (1930) suggest that giant fibre branches extend into the neuropile and it is conceivable that these may enter the central ends of the segmental nerve.

The input side of the reflex arc is also subject to rapid accommodation. Stimulation of a segmental nerve through the body wall gives rise to impulses in the giant fibre in a 1:1 ratio. The sensory fibre:giant fibre synapse accommodates after about one-third of a second. This means that at a stimulation frequency of 3/sec only one or two impulses can be monitored from the giant, at 22/sec some eight impulses are obtained (Fig. 49). Laverack (1960, unpublished) has found that whilst this is generally true for the median giant in response also to chemical stimulation, it is not true for the lateral giants since impulses can be recorded for some seconds after application of chemicals, and indeed the initial rapid volley of giant potentials also lasts longer than one-third of a second. It is possible that a somewhat different set of conditions is encountered here, for flooding the body wall with chemical solutions leads to stimulation of many sense organs situated in the body wall, and as such organs do not adapt anywhere near as rapidly as the giant fibres, and recovery of the sensory:giant fibre synapse is very quick (Roberts, 1960) it is likely that continual external stimulation maintains impulses in the sensory nerves for long enough to re-excite the giant fibre at different sites after a recovery period. The rapid exhaustion of the giant fibre reflex therefore seems to be a function of the giant to motor fibre synapse, since recovery periods are not followed by renewed conduction of the impulse. The sensory:giant synapse also fatigues quickly, but recovers quickly as well, the neuromuscular junction conducts impulses for long periods. Different types of stimulation, however, can lead to differential exhaustion of the giant fibre reflex according to Kuenzer (1958). If an earthworm twitch response is completely exhausted by mechanical means such that even the heaviest tactile stimulation leads to no giant fibre activity, it is still possible to obtain twitching by electrical stimulation. Perhaps this indicates that the sensory:giant synapse is after all the more important junction. It becomes less easy to exhaust the rear segments after removal of the sub-oesophageal ganglion or severance of the ventral nerve cord in front of the clitellum. If the ventral nerve cord is cut behind the clitellum it becomes less easy to stimulate all the segments.

## Control of Gut Movements

The body of the earthworm may for simplicity be considered as a tube within a tube. And in some ways the physiology of each of the tubes is similar. Both show peristaltic movements, and both appear to be under nervous and humoral control.

The intestinal tract of *Lumbricus* receives a double set of nerves, analogous to those of the vertebrate gut. One set of nerves arises from the circum-oesophageal nerve ring, and forms a plexus that runs between the mucus membranes and the muscle layers of the intestine. The second set of nerves comes from the ventral nerve cord running up via the septum to the gut.

Wu (1939a) considered these two nerve supplies to be antagonistic in function. Excitation of those arising in the ventral nerve cord led to a decrease of the motility of the gut wall, whilst stimulation of the nerves from the oesophageal commissures is excitatory to the gut. Millott (1943 a, b) questions this simplified view, considering that the septal nerves contain both excitor and inhibitor fibres, and that the function of the oesophageal nerves is not clear. Millott (1944) also noted that stimulation of the ventral nerve cord not only affected the tone of the gut but also increased the secretion of protease enzyme within the gut lumen. He adduced evidence that the nervous pathway involved in this system passes from the ventral nerve cord to the peritoneum applied to the body wall inner surface. The nervous pathways then ran through the ventral part of the intersegmental septum and thence to the plexus of the gut. Quite why such secretory-stimulating nerves should come via the body wall is not apparent, especially as they seem to be located more on the internal surface of the muscle layers than deeper within the body wall.

Pharmacological evidence indicates that despite the confusion regarding innervation the system is undoubtedly a double one. The muscles of the intestine of earthworms contracts under the influence of acetylcholine (Wu, 1939; Millott, 1943 a, b; Ambache *et al.* 1945), and the sensitivity of the gut is somewhat increased by eserine. The response is inhibited by atropine treatment. A cholinesterase specific enzyme has been demonstrated in the gut (Millott, 1943a) and presumably is the enzyme involved in breaking down internally produced acetylcholine.

The synthesis of acetylcholine by the gut nerves is shown by the experiments of Ambache, Dixon and Wright (1945). The normal peristaltic activity of the isolated intestine of *Lumbricus* and *Allolobophora* is diminished by cooling to 0–7 °C, but the gut still reacts to the addition of acetylcholine. The cooling action interferes with normal acetylcholine production and thus inhibits the intrinsic nervous activity of the intestine. Gut preparations kept in saline in the presence of eserine at 17–20 °C release into the medium substances capable of causing contraction of other similar preparations. Alkali destroys this activity.

Inhibition of the muscular movements of the intestine of *Lumbricus* is also an active process (Wu, 1939 a, b; Millott, 1943 a, b). The responses of the gut to the addition of adrenaline depends (a) upon which part of the gut is considered, and (b) on the concentration of adrenaline used. The anterior portions of the gut, buccal cavity, pharynx and oesophagus are stimulated to contract by adrenaline in all concentrations. On the posterior portions of crop, gizzard and intestine adrenaline has two actions, one inhibitory and one stimulatory depending upon the amount added. Acetylcholine-induced contractions can be counteracted by large dose levels of adrenaline.

After cooling the gut adrenaline has no effect upon it and Ambache *et al.* (1945) consider that this indicates that the usual action of adrenaline is to potentiate the action of acetylcholine. Ephedrin, used as an antagonist to adrenaline in vertebrates, slightly increases the actions of adrenaline in the earthworm gut. Ergotoxin abolishes the action. Millott (1943b) injected adrenaline, acetylcholine, tyramine and ephedrine into the blood stream of *L. terrestris* through the pseudohearts, and found that the actions of the nervous system could be mimicked. He concluded that the nerves controlling the tone of the gut were cholinergic and adrenergic, and that the two systems are antagonistic, not potentiating. Wu (1939b) decided that the nerves arising in the septa were the adrenergic variety, since their action was abolished by ergotoxin.

## Transmitter Substances

The identification of possible transmitter substances in oligochaetes at present depends upon the comparative pharmacological effects of organ extracts and synthetic substances. By means of studying muscle responses and antagonistic actions Umrath (1952) showed that nerve extracts of annelids, including *Lumbricus*, *Lumbriculus* and *Tubifex* contain acetylcholine. Enders (1952) also demonstrated acetylcholine in extracts of the nervous system of *Lumbricus*. Adrenaline is also present (Enders, 1952) although Ostlund (1954) was unable to obtain enough material to estimate the adrenaline content by paper chromatography. Unpublished experiments (Laverack, 1958), however, have demonstrated that 5 hydroxytryptamine has very similar intestinal excitatory effects to those described above, and Welsh and Moorhead (1960) have now supplied evidence that 5HT is present at a concentration of $10.4 \mu g/g$ fresh wt. No physiological function is at present known for it.

The evidence for the presence of acetylcholine seems fairly conclusive and is as follows:

(1) Stimulation of the gut nerves causes a rise in tone and motility of the gut wall, and is mimicked by the addition of synthetic acetylcholine.

(2) These actions are both depressed by atropine.

(3) They are also both stimulated by eserine.

(4) Cholinesterase is present.

(5) Acetylcholine-like substances are released in the presence of eserine which inhibits cholinesterase.

(6) Acetylcholine accelerates, and then stops in systole, the beat of the pseudohearts of *Lumbricus* (Prosser and Zimmerman, 1943).

(7) Nicotine abolishes nerve effects on the body wall.

The probability of adrenaline being present is indicated thus:

(1) Chromaffin cells are present in the ganglia of the ventral nerve cord (Gaskell, 1914).

(2) The rate of contraction of the blood vessels is accelerated by adrenaline.

(3) Adrenaline sometimes antagonizes and sometimes potentiates the action of acetylcholine on the gut musculature.

(4) Extracts of the worm ventral nerve cord cause increases in blood pressure when injected into cats (Ostlund, 1954).

(5) Nerve extracts also relax fowl rectal caecum preparations (Ostlund, 1954).

(6) Adrenaline effects on the gut are increased by ephedrine, and reduced by ergotoxin.

(7) Although the evidence is inconclusive Blaschko and Himms (1953) believe the presence of an amine oxidase in the ventral nerve cord is very likely.

Von Euler (1948) and Ostlund (1954) both suggest, however, that adrenaline is not present alone, but is accompanied by noradrenaline. Ostlund could not confirm this by chromatography because of lack of materials. Von Euler (1961), however, by use of a fluorimetric technique has shown that adrenaline is present to the amount of 0·003 $\mu$g/g, and noradrenaline 0·015 $\mu$g/g in *Lumbricus terrestris*.

*Giant Fibres*

In transverse sections of the nerve cord of the oligochaetes it is to be noted that dorsally there lie three large areas which are in fact huge axons. These are the giant fibres. They extend the whole length of the body from the cerebral ganglion to the pygidium, and in each segment several large cells send their axons into these fibres. There is a median large fibre, flanked by two smaller lateral fibres. In each segment two pairs of giant cells send processes to the median fibre, and the lateral fibres are connected to one another by transverse fibres. Branches are believed to be given to the motor neurones in each segmental nerve. Two rather smaller giant fibres lie ventrally in the nerve cord, but they remain unknown physiologically (Stough, 1926, 1930; Smallwood and Holmes, 1927).

These giant fibres are concerned in the rapid withdrawal or "startle" response of the earthworm. This has been shown many times by experiments involving cutting, regeneration, embryonic development and blocking with a number of depressant drugs (Prosser, 1933). Stough (1930) believed that the giant fibres were polarized and conducted in one direction only. By cutting the median fibre he found that anterior stimulation initiated muscular responses as far back as the cut but not beyond. Posterior stimulation caused the whole animal to contract. If he severed the two lateral fibres anterior stimulation invoked contraction of the whole animal, but posterior stimulation produced contraction only up to the lesion. He therefore concluded that the median giant fibre conducts from anterior to posterior and that the lateral fibres conduct in the reverse direction.

The first electrophysiological experiments carried out on these fibres (Eccles, Granit and Young, 1933) showed that conduction occurs equally well in both directions in both median and lateral giant fibres. When the fibres were stimulated electrically there was an all-or-none response above threshold that gave rise to two impulses, one travelling faster than the other. The faster of these is that of the median giant, the slower is in the lateral fibres. The observation that only one impulse occurs in the lateral giants indicates that the transverse connections are physiological as well as anatomical.

This result posed a problem that even now has not been properly resolved. Stough (1926, 1930) considered that polarization of the giant fibres was the function of the oblique septa which cross the fibres at each segment. They slope antero-posteriorly in the median giant, and postero-anteriorly in the lateral giants. They are histologically complete and divide the giant fibres into a series of blocks of tissue which are segmentally arranged. These synapses, if such they are, do not conduct in one direction as was shown by Eccles, Granit and Young (1933). This is an unusual situation and even more unusual is the fact that the delay time taken for an impulse to cross from one side of the synapse to the other is very short indeed and considerably less than normal vertebrate synapses, since the delay involved in crossing 50–100 synapses is only 5 msec (Bullock, 1945).

The action potentials of the giant fibres are extremely large and Rushton and Barlow (1943) were able to record them from the external surface of the animal without any need for dissection. The speed of conduction has been estimated as 17–25 m/sec in the median giant, (Eccles, *et al.*, 1933; Bullock, 1945; Rushton, 1945) and 7–12 m/sec in the lateral giants, at a temperature of 10–12 °C.

The conduction velocity of these fibres can be changed by alteration in the temperature. A rate of 18·8 m/sec in the median giant at 22·3 °C can fall to 10·8 m/sec at 8·2 °C. In the lateral giant fibres the comparable rates were 9·6 m/sec and 5·7 m/sec respectively, a decrease of similar proportions. The mean $Q_{10}$ is 1·52 for the conduction velocity of the median giant, and 1·46 for the lateral giants. The giant fibres continue to conduct impulses over a temperature range of −5 °C to + 36 °C, both of which temperatures are beyond the survival limits of the intact animal (Hogben and Kirk, 1944). Temperatures above +40 °C or below −5 °C produce irreversible changes and conduction ceases, failing in the median fibre before the laterals in the majority of cases (Turner, 1955).

The early results of Eccles *et al.* (1933) on the electrical properties of the giant fibres were confirmed and expanded by Bullock (1945). Two waves representing the median and lateral giant fibres were observed after stimulation. Rushton (1945) showed that the fast wave is characteristic of the median giant by applying a pressure block to this fibre and noting the disappearance of the fast response; he dealt similarly with the lateral fibres and noted the absence of the slow wave. On cutting the ventral nerve cord laterally, alternately left and right at intervals of ten segments he found that both waves were still functioning but the conduction time of the slower wave increased by 0·8 msec for each section after the first, and the potentials in each lateral get out of phase with one another despite the cross connections between them. But if a sufficiently long length of nerve is examined, say over 100 segments, the wave which should be slowed down in fact is able to catch up and fuse with its fellow.

Bullock (1945) was able to show that although the giant fibres may not be polarized and can conduct impulses in either direction they are unable to do so while in the intact animal. Mechanical sensory stimulation of the anterior segments, back to and just beyond the clitellum, gives rise to potentials only in the median giant. Stimulation posterior to this region and as far back as the rear extremity elicits activity only in the lateral fibres (Fig. 50). Adey (1951) found that essentially the same results were obtainable from *Megascolex sp.*, the critical switch of conduction in the giant fibres occurring at about segment 65. This distinction of direction of conduction, not a function of the giant fibres themselves, can therefore only be due to the anatomical connections between the sensory input and the giants. Anterior sense organs make contact with the median giant and posterior sense organs contact the lateral giants (Rushton, 1946).

The difference in conduction rate between the two types of giant fibre is thought to be a function of the diameter of the axon, the

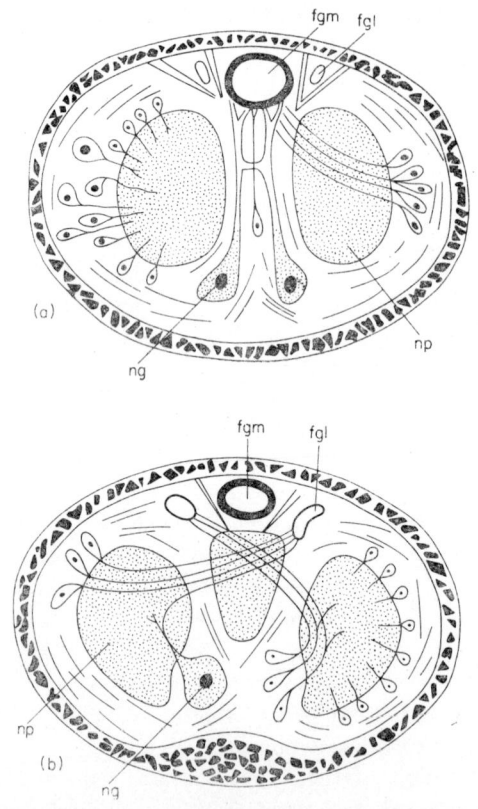

FIG. 50. Sensory input of median giant fibre (black) and lateral giant fibres. (a) in anterior sixty segments sensory neurons feed into median giant, (b) in segments to the rear of No. 60 the sensory nervous feed the lateral giants. fgl = lateral giant fibre; fgm = median giant fibre; ng = giant neurone; np = neuropile. The muscular envelope of the nerve cord is in black (from Grasse, 1959, after Adey, 1951).

median giant being larger and conducting faster. Adey (1951) examined a number of fibres in *Megascolex* for speed of conduction in relation to diameter and found a linear relationship, which did not, however, appear correlated with the degree of stretching of the body. In such an extensible animal as the earthworm the nervous system must necessarily undergo a considerable amount of stretching and Bullock (1945) believed that this stretching must alter the rates of conduction. Bullock, Cohen and Faulstick (1950) later modified this opinion and state that the conduction velocity remains constant regardless of the degree of stretching to which the nerve cord is subjected.

Supernormal rates of conduction within the giant fibres of *L. terrestris* are noted when the giant fibres are conditioned by a few (2–5) previous impulses. This is termed facilitation of conduction rate by Bullock (1951). The phenomenon reaches a maximum after a few impulses, being independent of stimulus frequency. It begins within a few msec of the absolute refractory period, can rise to 10–20% above normal conduction rates and declines slowly, being still appreciable after 100–200 msec. Fatigue occurs with continuing stimulation. This facilitation does not seem to be due simply to a phase of supernormal excitability after absolute refractoriness since high rates can be obtained when excitability is low and *vice versa*.

The most recent work on the electrical properties of the earthworm giant fibres is that of Kao (1956) and Kao and Grundfest (1957) in which the application of microelectrode implants have produced very interesting results.

The resting potential of these axons is about −70 mV. The size of the action potentials varies from 80–100 mV, reversing the sign of the potential by up to 30 mV in some cases. The spike lasts for only 1 msec at 20 °C, there being no hyperpolarization phase at the start, and only a brief depolarization at the end of the action potential. The separation of median and lateral giant fibres shown by Bullock (1945) and Rushton (1945) is also confirmed since completely different sets of spikes can be obtained from the two systems (Fig. 51).

Intracellular recordings using microelectrodes have also been carried out on the lateral giant fibres by Wilson (1961). The results confirm Rushton (1945) in the presence of electrical connections between the lateral giant fibres on either side of the body. There is considerable attenuation of spike size in the passage from one side to the other. Impulses in one fibre may be blocked by anodal stimulation, but there is no delay in transmission since the bridge to the second lateral conducts rapidly and the unblocked fibre determines conduction velocity in both fibres.

Normally single stimulation by external electrodes causes one

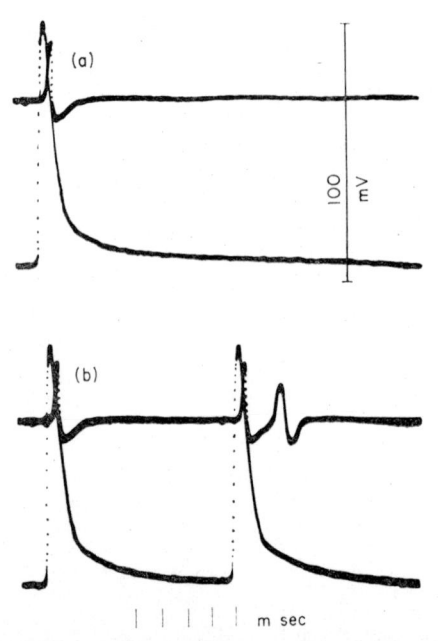

Fig. 51. Correlation of internally and externally recorded spikes. Microelectrode impaled median giant axon, between a pair of external electrodes. External trace is zero reference for resting potential; (a) weak stimulus, threshold for median giant axon elicited diphasic responses on external trace, corresponding to internally recorded spike of that fibre; (b) second, stronger stimulus excited also lateral giant axons whose spike appears as later diphasic response. Activity of these fibres does not distort internally recorded spike of median giant axon (from Kao and Grundfest, 1957).

spike to occur, but occasionally repetitive firing is observed. This has been noted previously by Amassian and Floyd (1946). The first spike rises rapidly but the later ones are preceded by a prepotential that rises slowly. The cause and nature of this repetitive discharge was studied by applying brief shocks in rapid succession.

After a threshold shock a second stimulus was applied in the relative refractory period of the first. As a consequence the potential remained high and numerous small peaks were observed and later another spike potential occurs. The local, non-propagating nature, low amplitude, frequency of repetition, and additiveness of the small potentials suggest that they are synaptic, developing at the junction of small nerve fibres with the giant axons. These post-synaptic potentials are then responsible for re-exciting the fibre, and may explain the after-discharge phenomena of reflex arcs in the earthworms (Collier, 1938, 1939 a, b) previously explained by the production of humoral transmitters.

The post-synaptic nature of these potentials rests on six points:

(1) They are not artefacts from neighbouring fibres since the spike of the lateral giants does not cause potential changes in the median fibre.

(2) They are independent of the spike potential since they occur during the absolute refractory period and appear upon the falling phase of the spike.

(3) They appear at sites which do not generate spikes.

(4) They are graded and they summate.

(5) They arise at various sites on the axon, often far from the site of stimulation.

(6) They give rise to spikes that may be initiated at different sites of the axon (Kao and Grundfest, 1957).

The histological work of Smallwood (1926), Stough (1930) and others makes no mention of synaptic endings impinging upon the surface of the giant fibres, but branches of these axons are known to ramify into the neuropile that comprises the rest of the ventral nerve cord, and it is in this tangled mass of fibres that the connections between giants and sensory and motor fibres must occur, and where the synaptic potentials mentioned above have their origin. It is suggested that these giant fibre branches may be afferent or efferent (Kao and Grundfest, 1957).

*The Septa*

As figured by Stough (1930) the giant fibres are divided into discrete sections by membrane partitions that cross the axon completely sloping either forward or backward, according to the fibre studied. Later microscopical studies showed that the nerve sheath extended along the surfaces of these discontinuities. This sheath material was thought to be myelinated and to have radially orientated lipid molecules (Taylor, 1940). The presence of a definite

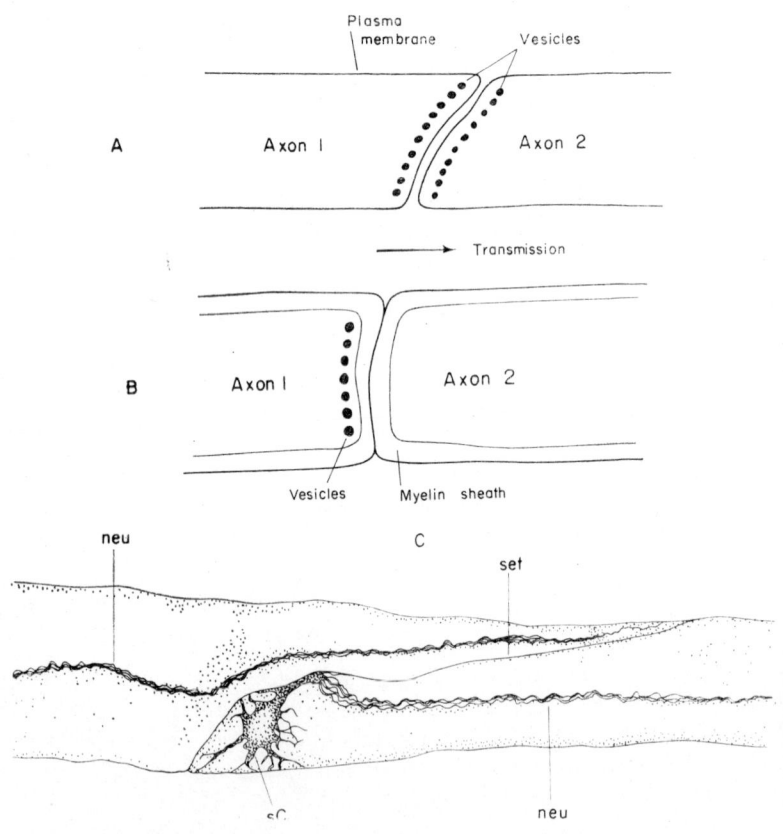

FIG. 52. Structure of intersegmental septum. A. after Hama, 1959, B. after Issidorides, 1957; de Robertis and Bennett, 1955. C. Septum of median giant fibre, set = septum, sc = small ganglion cells, neu = neurofibrils (from Ogawa, 1938).

membrane structure was described after electron microscope studies (de Robertis and Bennett, 1955 a, b; Issidorides, 1956) though there are some disagreements over exact details. Numerous small vesicles were figured on the pre-synaptic side of the septa, and in the neuropile extrusion of these vesicles through small pores in the axoplasmic membrane was noted. This finding of vesicles only on the pre-synaptic side suggested that some degree of polarization is present, despite the electrophysiological evidence to the contrary, but Hama (1959) demonstrated that such vesicles are to be found on both sides of the septa, pre-synaptic and post-synaptic. Thus on histological grounds there is no evidence for polarization. Hama (1959) also considers that the myelin-like lamellae figured by earlier workers are absent, and indeed the layers of the septa are extremely thin (Fig. 52).

Although the discontinuities in the axons, known as septa, are composed of thin layers and appear to have a similar construction on both sides there is little knowledge yet of their effect upon the

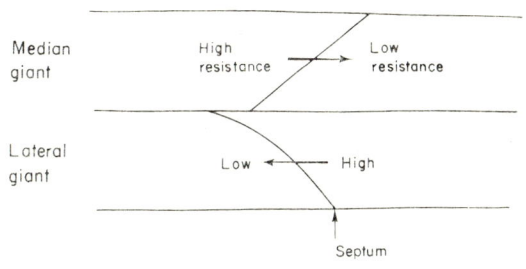

Fig. 53. Possible explanation of functioning of septa in giant fibres of *L. terrestris* (based on ideas of Kao and Grundfest, 1957).

transmission of the impulse along the axon. They must allow the conduction across the gap to occur with the utmost rapidity and pose little in the shape of a barrier to the potential.

A possible explanation of this phenomenon comes from Kao and Grundfest (1957) by analogy with another giant fibre system, that of the crayfish. Because of the relatively short length of the individual segments electrotonic spread can account for the spread along the individual sections. The septa are considered to be ephapses, not true synapses, that is the simple physical adjunction of nerve fibres from two segments. In the crayfish there is a low resistance in one direction and a high resistance in the other along the length of a giant fibre so that conduction proceeds in one direction only. If a similar set of resistance conditions are present in the earthworm it may account for functional polarity in the living animal, although the initial input is probably governed by the sensory connections (Fig. 53).

*Sense Organs and Sensitivity*

As late as 1955 Hodgson, reviewing the gaps in knowledge of invertebrate chemoreception reported that nothing was known of the sense responses of *Lumbricus*. Although the histology of the sense organs of *Lumbricus* had been described in detail by Langdon (1895) and Smallwood (1926) "experimental verification of the presumed chemosensory function is lacking".

Prosser (1935), however, has given us some information on the sensitivity of earthworms to a number of chemical substances as well as to light and touch. The segmental nerves of the earthworm are mixed nerves, there being no division into sensory and motor nerves amongst the segmental nerves. Large spike potentials are obtained in response to tactile stimulation of the epidermis with a fine needle, and also in response to pulling and stretching the body wall. This latter observation indicates that stretch receptors are indeed present as suggested by earlier experiments on locomotion.

The touch receptors of the earthworm segments serve discrete areas on the surface of the body. Each of the three nerves in each segment covers particular regions of the segment of origin, and also the segment immediately before and after. The anterior nerve in each segment has a greater receptive area on the segment ahead, whilst the posterior nerve has a greater sensory area on the segment behind. These results have been confirmed by Laverack (1960b).

Light also evokes potentials in the segmental nerves. This response is only noticeable when the earthworm has undergone a considerable period of dark adaptation, illumination after 15 minutes darkness causing no potentials to appear, after 30 minutes, such treatment a small burst is noted, and after an hour of dark adaptation a large volley is seen.

Prosser (1935) found no nervous activity was elicited by the application of sucrose solutions to the body wall, but obtained records after stimulation with HCl, NaOH, and NaCl. N/50 HCl provoked a short burst of potentials. N/20 HCl gave a larger burst and N/10 killed the sense organs of the skin.

These results have since been extended by Laverack (1960b)

who found that 0·1 M NaCl was the threshold value necessary to fire salt-sensitive fibres, the initial frequency of the impulses increasing as the strength of the stimulus increased. Sucrose and glucose do not affect the sense organs of the body segments, confirming the observations of Prosser (1935b), but the prostomial nerves contain a few fibres that respond to glycerol and sucrose, though not to glucose. The prostomial nerves are also sensitive to quinine at a concentration $10^{-5}$ M to $10^{-4}$ M.

The importance of the hydrogen ion in governing the distribution of earthworms has been suspected for some time (Bodenheimer, 1935; Hurwitz, 1910; Arrhenius, 1921; Satchell, 1956).

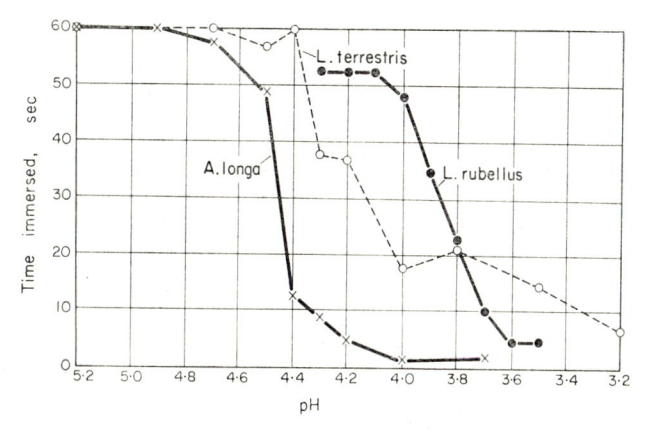

Fig. 54. Graph to show the average time taken to withdraw the prostomium from acid pH solutions in three species of earthworms × *A. longa*; ○ *L. terrestris*; ● *L. rubellus* (from Laverack, 1961a).

Different earthworm species have slightly different habitats in nature, and this may be reflected by differential sensitivity to acidity. By applying buffer solutions to the body wall of these species *A. longa* was shown to have a threshold response at pH 4·5, *L. terrestris* at pH 4·2–4·1, and *L. rubellus* at pH 3·8 (Laverack, 1961a) (Fig. 54, 55).

*Motor Nerves*

Very little is known of the functioning of the motor nerves.

Prosser (1935) noted that the segmental pattern of the motor fibres is very similar to that of the sensory fibres, extending to the segments in front of, and behind, the segment of origin, but the effects of stimulation are not so pronounced in the case of the motor axons. The section of one ganglion in the nerve cord delayed the passage of peristaltic waves along the body, and the complete removal of one ganglion stopped peristalsis. Peristalsis only passes

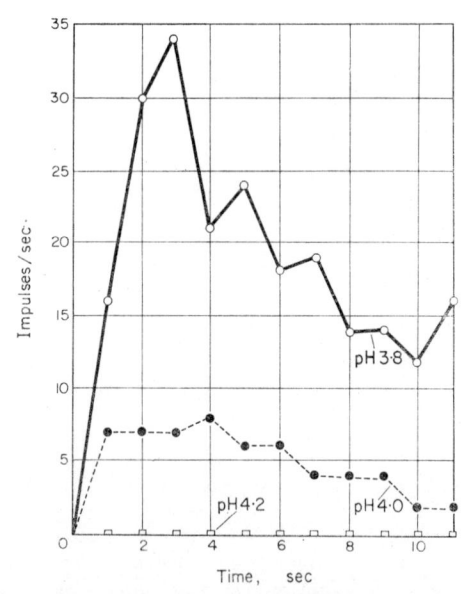

FIG. 55. A single nerve unit of *L. terrestris* which gave potentials in response to pH 4·0 (●) and pH 3·8 (○), but not to pH 4·2 (□) (from Laverack, 1961a).

again when one segment is allowed to pull on another, so the peripheral plexus and stretch responses are unable to conduct impulses across more than one segment.

This has received support from Miller and Ting (1949) who consider the epidermal net to be segmental in arrangement, peristalsis spreading through one segment, but requiring further stimulation of receptors in the succeeding segment before progressing along the body.

*Summary*

Peristaltic movement in the body wall of the earthworm is coordinated by interaction of central and peripheral nervous activity. Conduction of peristaltic nerves can occur via the ventral nerve cord, but in the case of interruption of this cord can be transmitted

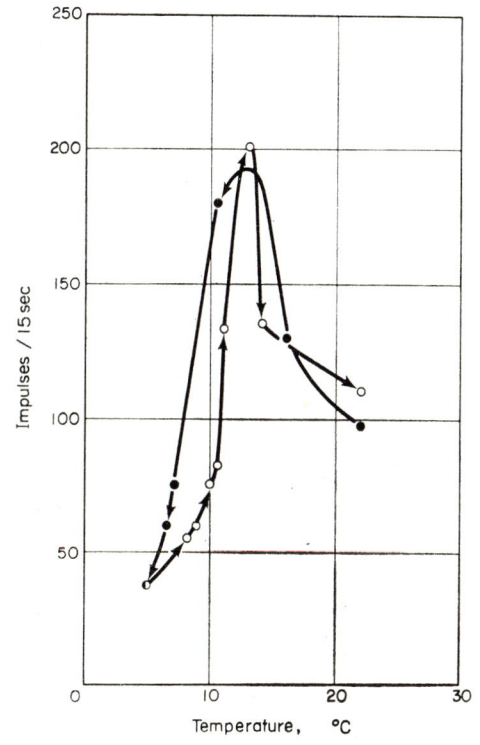

Fig. 56. Graph showing steady rate curve of nervous activity for a range of temperatures. The number of impulses rises towards an optimum at 13 °C and then falls as the temperature increases further (from Laverack, 1961b).

by tension on the body wall in which stretch receptors are probably present. Rhythms of electrical activity have been described in the ventral nerve cord with a periodicity similar to that of peristalsis. The properties of the components of the reflex arc, sensory, central association and motor, are partly known but much work has centred on the three giant association fibres of the earthworm. These are quick conducting pathways running the whole length of the oligochaete body. Sensory input occurs into the median giant anteriorly and to the lateral fibres posteriorly. They contain septa which are not true synapses and conduction can proceed in either direction across them although a functional polarity in the animal is probable. Transmitter substances represented in the central nervous system are acetylcholine, probably adrenaline and noradrenaline and possibly 5HT. The properties of some sense organs have been described.

CHAPTER XI

# BEHAVIOUR

THE work of Lorenz, Tinbergen, Thorpe, Lack and others in recent years has thrown much light and insight into the behavioural characteristics of vertebrates, particularly mammals and birds. Observations by Pantin, Blest, Wells, etc. have provided similar observations for a few invertebrate species but on the whole the impetus of modern behavioural observations has been towards elucidation of vertebrate systems.

The life of many invertebrates may seem simple at first glance but as the analysis by Wells and co-workers of the behaviour of *Arenicola* shows, many complex processes are carried out. No one has as yet subjected *Lumbricus* or any other oligochaete to such intensive investigation but none the less a number of facts can be reviewed here.

Earthworms and other oligochaetes tend to be rather secretive animals, nocturnal in habit and easily irritated into hasty removal. As is the case with many other characteristics of the oligochaetes most of our knowledge is confined to the earthworm species. A few behavioural facts are known with regard to *Tubifex*, but these are dealt with elsewhere (Chapter VII on respiration).

The behaviour of earthworms is influenced, as is that of any other animal, by the changing conditions of the environment which surrounds it. The soil in which earthworms live may seem a fairly rigid and unchanging medium to inhabit, but even here such properties as the acidity of the substrate, the water content, food content, leaf cover and even gas concentrations may have an effect upon the activity and behaviour of earthworms. Most of these factors, and the various changes they undergo, will be detected by the earthworm through the medium of the sense organs. But as we shall see information regarding these sense organs is at present rather scanty. The morphological and anatomical relations of the sense organs in the skin were described as long ago as 1895 by Langdon, but only within the last few years have modern electrophysiological techniques been applied to a study of the actual sensitivity of such organs.

Prosser (1933) has followed the development of the characteristic behaviour of the embryos of *E. foetida* and has shown that as the structural differentiation of the body proceeds so does the activity and response to external stimuli become more complex. The first responses to tactile stimulation are local contractions of the body wall which are correlated with the appearance of cells of the circular and longitudinal muscles. Later on flexion of the anterior end in response to touch requires not only the presence of the muscle layers, but also fibres in the CNS and lateral nerves in the sensitive area. The more complex behaviour of extension, peristalsis, exploration and co-ordination of setae only occurs when the circular, longitudinal and transverse fibres of the nervous system are complete. The response to illumination of the anterior end is correlated with the development of prostomial nerves bearing enlargements which probably contain photoreceptors, and with the appearance of central nervous connections. This is followed by the development of swift end-to-end contractions mediated via giant fibres.

The macroscopic behaviour of *Eisenia* (*Allolobophora*) *foetida* was studied by Smith (1902). This species is a well-known inhabitant of piles of rotting vegetation and manure where the organic content is high, and very often the temperature also is higher than in the adjacent soil. The temperature in such a pile of rubbish can sometimes rise as high as 20 °C. Smith found that if she put *E. foetida* on a glass plate which she proceeded to warm gradually the worms showed no response as the temperature rose from 20 to 30 °C, but that if the temperature continued to rise above 30 °C the animals moved away from the source of heat. Heat death occurred at a temperature of 36–40 °C. This compares with a heat death temperature of $29 \pm 0.5$ °C when allied with desiccation (Hogben and Kirk, 1944). It seems unlikely from these observations that environmental temperatures play much part in governing the activity of this species since the temperatures encountered in field conditions rarely rise high enough to affect overt activity. At such times as intense summer heat may occur it is always possible for earthworms to burrow down into the cooler depths of the soil, and thereby maintain themselves in the optimal conditions. In a similar way the frozen upper layers of the soil may be avoided during wintry conditions.

The factor which attracts *E. foetida* into a heap of manure does not seem, therefore, to be the heat generated in such a site. Smith (1902) found that if *E. foetida* is released anywhere near rotting substances they take no notice of it whilst they are any distance from it. Thus it seems that the animals can neither smell nor detect heat from a distance. The lack of response of the nervous system to distant heat has been confirmed by Prosser (1935). This observation holds true even when the worms pass within a millimetre or two of the material. If, however, the prostomium should happen to come into contact with the manure the worm immediately begins to burrow inwards, suggesting that chemo-reception is the important factor here. The lack of sensitivity to manure from a distance suggests that the earthworm is not capable of reacting to odorous substances, but it is found to react to such things as cedar oil, xylol, ether and turpentine without making physical contact with any of them. Such observations would repay further investigation.

One of the best-known behavioural responses of earthworms is that of thigmotaxis. When placed upon the surface of the ground, or on a sheet of paper or glass, the animals are usually very active, but if they should find their way into a crevice or crack which means that the sides of the body are in contact with the substrate then very often the individual ceases to move, remaining quietly in this artificial burrow. This thigmotactic response, the reaction to surrounding contact, overrides the usually strongly negative action to light noted upon illuminating the animal.

## Light

Earthworms, unlike some of their polychaete relations, do not possess eyes as discrete structures although the oligochaete group Naididae do have a pair of eye spots, about the physiological function of which nothing is known. The earthworms do have single sensory cells which contain a lens-like structure lying in the epidermis and dermis, particularly in the prostomium, which are believed to be light-sensitive (see Stephenson, 1930).

The behavioural reactions of earthworms to light depends to a certain extent upon the immediate past history of the animals. For example Hess (1924) kept *L. terrestris* in the dark for some hours, and then exposed either the right or left side of the body to a beam of light, recording the movements of the anterior end. In all except the weakest light (0·0018 metre-candles) the animals

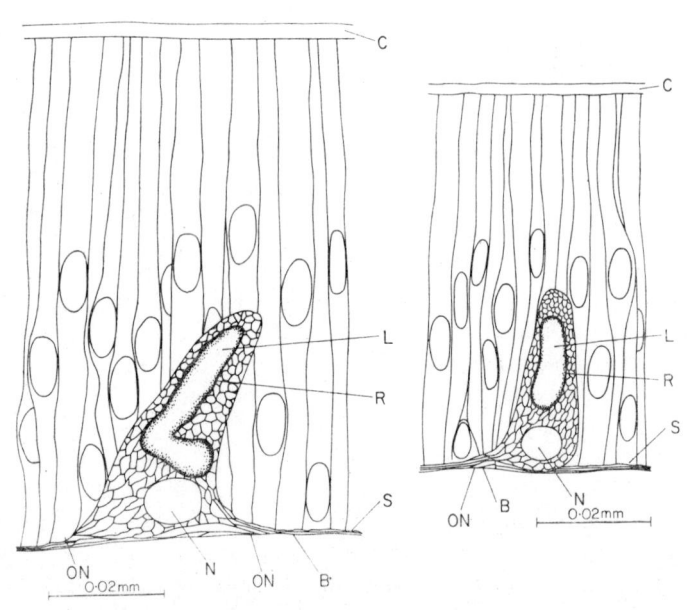

FIG. 57. Epidermis of prostomium, showing photoreceptor cells with two optic nerves. B = basement membrane; C = cuticle; L = lense; N = nucleus; ON = optic nerve; R = retinella; S = subepidermal nerve plexus (from Hess, 1925).

responded by moving away from the light; in the weakest light the earthworms were photopositive. In other words earthworms crawl towards very dim light, but away from more intense sources. If, however, *L. terrestris* is subjected to long periods of moderate illumination they often do not react at all to a sudden increase in the intensity of the light (Hess, 1924), due perhaps to adaptation or to complete saturation of the light receptors. Long periods of dim illumination sensitize *L. terrestris* so that it reacts positively to more intense light than when it is dark-adapted.

Another species of earthworm, *Pheretima agrestis*, shows only one response to light, avoidance. The photonegativity becomes more pronounced as the stimulus intensity increases, and although adaptation occurs in dim light so that the threshold illumination necessary to stimulate the animal rises, the response remains photonegative. Other types of stimulation, mechanical and thermal,

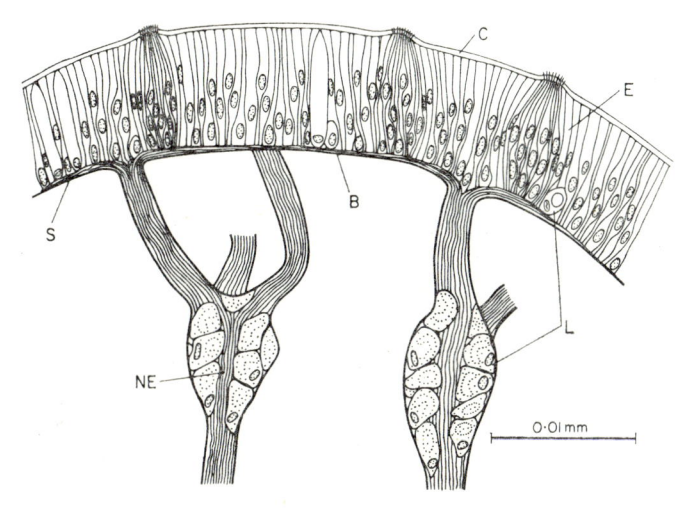

Fig. 58. Nerve enlargements carrying photoreceptor cells in prostomium. E = epidermis; L = photoreceptor cells; NE = nerve enlargements; other letters as in Fig. 57 (from Hess, 1925).

for example, decrease the sensitivity of the light response (Howell, 1939).

The quality (wavelength) of the light, as well as the quantity (intensity), also plays a part in the light stimulation of earthworms. Bretnall (1927) passed light through a prism and projected the resulting spectrum onto a sheet of white paper. He then dropped worms on the paper and noted that all except one moved away from the blue colours and travelled across the spectrum to pass out at the red end. The dissenting individual lay full length in the green band. It would seem that red, rather than being an attractive waveband, is non-stimulatory since earthworms can be studied comfortably when illuminated with red light and it has no apparent effect upon the normal activities of the animals. If blue light is flashed upon them they withdraw rapidly into their burrows (Walton, 1927).

This reaction to blue light is most pronounced at a wavelength of about 483 m$\mu$ (Mast, 1917). Although the intensity of the various wavelengths was not thought to affect these experiments it is none the less not possible to state definitely that earthworms are able to distinguish different wavelengths of light. Ultra-violet light may be lethal to earthworms, and it has been suggested that the many earthworms found lying dead upon the surface of the ground after rain have been killed by the u.v. of the sunlight (Merker and Bräunig, 1927).

*Regional sensitivity.* The experiments detailed above show that earthworms are capable of receiving light stimuli and will respond to such stimulation. Such sensitivity is not a uniform property of the body surface, however. If, instead of a large generalized source of light, a narrow pencil beam of light is projected onto the skin it can be shown that although light stimulation can be detected over the entire surface of the animal there are certain regional specializations. The prostomium is the most sensitive area of all, and the ability to react to light decreases as illumination passes rearwards along the body. The anterior end is more sensitive than the mid-region, but then sensitivity again rises towards the rear end (Hess, 1924, Howell, 1939). Within each individual segment the dorsal aspect is the most sensitive, and the edges of the segments at the intersegmental furrow are least reactive (Hess, 1924).

*Light and the Nervous System*

Interference with the nervous system in some form such as the removal of the cerebral ganglia, or the severance of the circumoesophageal commissures, causes at least a partial reversal of the normal responses to light. *L. terrestris* becomes photopositive in moderate light intensities, and *Pheretima agrestis* becomes positive to weak light, though it still remains photonegative in moderate intensities (Hess, 1924; Howell, 1939). Prosser (1934b) obtained similar results with *E. foetida* and also noted that depressant drugs and low temperatures lowered the responsiveness of earthworms to light. Section of the ventral nerve cord midway along the body, leaving the anterior portion under the control of the cerebral ganglion, leads to different responses of the anterior portion, which is photonegative, and the rear portion, which is photopositive (Hess, 1924). In all the reported cases (Hess, 1924; Howell, 1939; Prosser, 1934b; Nomura, 1926) the cerebral ganglion appears to be the ultimate controlling and co-ordinating centre.

Although it is obvious from above that the nervous system is intimately concerned with the reactions to light only one study has been made of the actual electrical activity occurring within the sensory nerves in response to such stimulation. Prosser (1935) recorded some small potentials from the sub-oesophageal ganglion which were of low frequency initially but increased to a maximum after 2 to 4 seconds. They were then overshadowed by proprioceptive impulses generated during contraction of the animal. Dark adaptation affects this response, no potentials being observed upon stimulation after 15 minutes in the dark, a small burst after 30 minutes, and a large burst after one hour. These nervous responses should be compared with the behavioural reactions mentioned above.

*Chemoreception*

It has been known since the time of Darwin (1881) that earthworms are able to distinguish between various food substances, demonstrating preferences for certain plants above others. The ability to select the material to be eaten seems to reside in the sense of taste. Chemoreception was indicated by Smith (1902) in her work on the reactions of *E. foetida* to chemical stimuli. This faculty of determining external chemical conditions probably plays a major role in the life of the earthworm. First it may be used as mentioned above, in the detection and gathering of food, secondly it may be used to give information regarding the other environmental conditions, such as soil acidity, and thirdly there is a possibility that it plays a part in mating by detecting the mucus secretions of other earthworms.

Earthworms usually come to the surface of the ground at night to graze upon leaves in the litter layer. They are able to draw leaves over the ground and down into the burrow. Not all leaves are utilized at the same rate, however, and selective feeding goes on.

This has been explained by Mangold (1951) on a basis of taste selection. He took pine needles, which could be easily drawn into burrows because of their shape, boiled them to remove their characteristic flavour which is not attractive to earthworms, and then coated them with gelatine containing finely powdered litter of various other leaves in suspension. From the rate at which such preparations were consumed Mangold concluded that when the litter was fresh, beech was the most favoured food plant, followed by maple, oak, horse-chestnut, lime, willow and false-acacia. If decaying leaves were used the order changed and became willow, false-acacia, oak, lime, beech, maple and horse-chestnut, indicating that probably the breakdown processes in the leaves change the acceptability of the leaves to the earthworms. There is also a possibility that maturation in the accumulation of polyphenols may be important (Satchell, personal communication).

Wittich (1953) in field experiments on naturally fallen and decayed litter found a different order of acceptance and explained this partly on a basis of the physical texture of the leaves, and partly on the presence of unpleasant substances in fresh leaves, such as beech, which break down as decomposition proceeds. As the inimical substances are removed from the leaves by leaching and decomposition so the order of leaf acceptance changes and the earthworms begin to graze on leaves ignored when freshly fallen. After chemical analysis of the leaves from various trees Wittich suggests that protein content, a measure of food value, is important to the order in which the leaves are eaten, but he does not suggest that the leaf proteins can be detected by the earthworms. It is possible that the protein content is correlated with some substance which can be detected by chemoreceptors.

Mangold (1953) extended his observations on the plants eaten by earthworms to include synthetic chemicals such as alkaloids, sugars and acids. He used the same experimental method as before including the known chemicals in the gelatine covering pine needles. Alkaloid substances are not taken by earthworms above a certain concentration, 50% acceptance being noted at a concentration of 0·01 g in 20 g gelatine, but concentration greater than this was greeted by complete rejection. Low concentrations of a number of acids often found in plant materials, e.g. phosphoric, tartaric, citric, oxalic and malic acids, were accepted but not high concentrations of these substances. Glucose and saccharose did not invoke consistent responses, sometimes being taken, at other times not. It seems from these results, however, that the food preferences of earthworms can be explained by the chemical sensitivity of the organs of the body.

The ability of earthworms to detect acids is of considerable interest in another direction. It has been noted (Bodenheimer, 1934; Satchell, 1956) that many species of earthworms are absent from acid soils. Each species shows a particular tolerance to acidity, but has a lower ecological pH limit below which, apart from a few individuals, it does not spread. It is also possible that other factors may be involved in this avoidance of acid sites, for example the low calcium levels allied to low pH may cause starvation effects (Jefferson, 1956) or give rise to osmotic abnormalities (Kopenhaver, 1937). The response of earthworms to acidity has been demonstrated a number of times experimentally.

Hurwitz (1910) was among the first to record what happens when earthworms come into contact with acid. He suspended *E. foetida* and allowed them to dip the prostomium into acid solutions, noting the time taken to withdraw the anterior end from the acid. The stronger the acid the shorter the time the prostomium spent immersed in it. HCl was the most effective at causing retraction, followed by nitric acid, sulphuric acid and acetic acid. The concentrations required bore little resemblance to field conditions. Neither did soils prepared by Arrhenius (1921) to have known pH values, by treating them with acids. He isolated *L. terrestris* and *Perichaeta indica* on these soils and found that survival was longest on soils near the neutral point. Bodenheimer (1934) using soils with a natural pH between pH 3 and 10 found that the length of survival of *Allolobophora samarigera* was at its greatest at around the neutral point of soil reaction. There was an optimal range over which the earthworms survived equally well, but it was pointed out that this optimal area may vary from species to species. Indeed although *A. samarigera* and many other earthworms are not found in the field at pH lower than 4·5, and indeed do not survive in such conditions, nevertheless *Lumbriculus variegatus* has been taken from acid sphagnum at pH 3·9, and *Bimastus eisenii* from peat moors at pH 3·7 (Satchell, 1956).

Another reaction to acid may be the explanation of the results of

Shiraishi (1954). This author showed that carbon dioxide gas can be detected by earthworms, if it is present in sufficient concentration. Animals moving along a tube can be induced to retrace their path when met headlong by a stream of carbon dioxide gas. This may offer an explanation of why earthworms leave their burrows after rain. If the open ends of the burrows are sealed by water during heavy rain a high concentration of carbon dioxide may build up, and at the same time the oxygen content of the air may fall. Carbon dioxide does not affect the respiration of earthworms (Johnson, 1942) but oxygen lack is reflected by a need for greater respiratory activity. But if the oxygen lack is only short-lasting earthworms are able to tolerate it by undergoing a period of anaerobic metabolism, repaying an oxygen debt later (Davis and Slater, 1928). Shiraishi (1954) suggests that the gaseous carbon dioxide can force the animals to leave the burrows, but this would need a very high concentration of $CO_2$ (see Chapter VII). This gas will dissolve in water, however, to form a weakly acid solution of carbonic acid, and as seen above acid sensitivity is a property of the skin sense organs of earthworms, and it may be this stimulus that causes vacation of burrows by earthworms, but this seems unlikely in species living in base-rich soils.

The sensory nervous foundation for these reactions are discussed in the section on the nervous system.

*Learning and Habituation*

Thorpe (1956) has recently reviewed very fully the processes and implications of learning and habituation throughout the animal kingdom. Such processes are shown by even the lowliest of animals, and the earthworm is no exception.

Among the earliest experiments designed to investigate the ability of earthworms to react to repeated stimuli and the consequent learned series of actions, were the well-known studies of Yerkes (1912 a, b). Specimens of *E. foetida* were put in the long arm of a T-shaped maze and allowed a choice in the cross arm of, to one side, a dark moist chamber in which the animal could burrow, and in the other, an area lined with sandpaper, beyond which the earthworm came into contact either with salt solution or an electric shock or both. Individual animals showed considerable daily fluctuation in the reaction to this problem, but it was usually found that after 20–100 tests the animal crawled only into the acceptable side. Habituation became so pronounced that Yerkes stated that after these trials there was an increased readiness to enter the apparatus and to desert it for the artificial burrow, an apparent "recognition" of the burrow with an increasing avoidance of sandpaper and stimulant, and less likelihood of a retracing of the original path along the stem of the T. The loss of the first five segments, including the cerebral ganglion, does not destroy these responses in a well-trained animal. As the regeneration of the head end proceeds, however, the reactions of the animal become more variable and less stereotyped. This continues until 2 months after regeneration is complete the acquired behaviour has disappeared, though the animal can be retrained again in about two weeks. The majority of these results were obtained with a single, rather exceptional worm, but Heck (1920) using *E. foetida*, *L. castaneus* and three species of *Allolobophora* obtained similar data. He also found that if the worms learnt one circuit which was then reversed it took significantly fewer trials for them to relearn the circuit. This effectively scotches the criticism that the turn to one side is a stereotyped response due to a few muscle blocks or a position habit. Evidently a certain amount of co-ordination in the central nervous system is required. Heck (1920) also showed that not only is the supra-oesophageal ganglion not essential for the maintenance of the response, but the animal can also learn as readily to navigate a maze without this organ as with it. It is possible to train worms to turn to the illuminated side in preference to the dark by continually shocking them in the dark arm of the T (Wherry and Sanders, 1941). The ability of earthworms to learn a maze may not be of great significance in the life of the animal for the number of times it comes up against such a choice repeatedly must be very few. Malek (1927) has demonstrated, however, that this system may be of some importance in another respect. *L. terrestris* (*L. herculeus*) will drag leaves to its burrow, but if the leaf turns out to be too large or stiff to get into the hole the attempt to draw it down ceases after ten to twelve trials. The observation that certain types of leaf are eaten in preference to others, and that physical shape and texture may be important in governing the choice has also been noted by Wittich (1953).

Yerkes (1912 a, b) remarked that there is a daily fluctuation in the performance of worms in a maze. This observation has been expanded by Arbit (1957) in a similar series of experiments carried out at 12-hour intervals. *L. terrestris* introduced into a T maze between the hours of 8 a.m. and mid-day took longer to acquire the habit of turning to the correct side than did worms trained between 8 p.m. and midnight. The latter group also needed less stimulation by touch or light to start them moving along the maze. Earthworms are evidently more "alert" during the night hours.

*Activity*

It is well known that earthworms come to the surface of the ground during the hours of darkness and some wander freely about feeding and mating. But anyone who has watched a bird on a lawn will not doubt that many earthworms are also to be found at the top of their burrow during the hours of daylight, even though they are photonegative to high intensities of light.

Studies have been made on the patterns of activity shown by earthworms under conditions of dark or low illumination. Baldwin (1917) watched earthworms and concluded that they were most active between 6 p.m. and midnight and Szymanski (1918) found that they were most active between 2 p.m. and midnight. More recently Ralph (1957) used actographs, dishes supported on knife edges, to provide a visual record of the motor activity of *L. terrestris*. The least activity was manifested between 7 and 11 a.m. with relatively high rates of movement before and after these times, reaching a peak in mid-afternoon and lasting through till early morning. This activity cycle did not reflect the diurnal rhythm of oxygen uptake, but no attempt was made to explain what happened physiologically. Harker (1960) has demonstrated a hormonally controlled "internal clock" in the cockroach which may explain the diurnal variation in the activity in this species. It is possible that in the earthworm there is another example of the same phenomenon, although no evidence is available. Neurosecretory cells, however, are present in the earthworm and all their functions cannot yet be known so that the postulation of an "internal clock" may not be completely fanciful.

*Summary*

The appearance of complex behaviour patterns has been

demonstrated to parallel the development of the nervous system. Earthworms show responses to light, particularly blue light, to heat, to bodily contact (thigmotaxis) and to touch. They show an ability to discern food materials and changes in them by their grazing on different leaves in litter on the surface of the ground. Acidity of soils affects colonization of sites and stimulation of acid-sensitive sense organs may account for migration from burrows of earthworms after rain. Activity can be modified by training in a maze. In normal conditions diurnal rhythms of locomotory activity are known.

# REFERENCES

ABDEL-FATTAH, R. F. (1955). The chloragogen tissue of earthworms and its relation to urea metabolism. *Proc. Egypt. Acad. Sci.* **10,** 36–50.

ABDEL-FATTAH, R. F. (1957). On the excretory substances in the urine and body fluids of earthworms. *Bull. Coll. Arts Sci., Baghdad,* **2,** 141–161.

ABELOOS, M. and AVEL, M. (1928). Un cas de périodicité du pouvoir régénérateur: La régénération de la queue chez les Lombriciens *Allolobophora terrestris* Sav. et *A. caliginosa* Sav. *C.R. Soc. Biol., Paris,* **99,** 737–739.

ACKERMANN, D. and KUTSCHER, F. (1922). Über die Extraktstoffe von *Lumbricus terrestris. Z. Biol.* **75,** 315–324.

ADEY, W. R. (1951). The nervous system of the earthworm *Megascolex. J. comp. Neurol.* **45,** 57–103.

ADOLPH, E. F. (1927). The regulation of volume and concentration in the body fluids of earthworms. *J. exp. Zool.* **47,** 31–62.

ADOLPH, E. F. (1943). *Physiological Regulations.* pp. 502. Jaques Cattell Press, Lancaster, Pa.

ALSTERBERG, G. (1922). Die respiratorischen Mechanismen der Tubificiden. *Acta Univ. lund.* N.F. AVD 2, **18,** 1–176.

AMASSIAN, V. E. and FLOYD, W. F. (1946). Repetitive discharge of giant nerve fibres of the earthworm. *Nature, Lond.* **157,** 412–413.

AMBACHE, N., DIXON, A. ST. J. and WRIGHT, E. A. (1945). Some observations on the physiology and pharmacology of the nerve endings in the crop and gizzard of the earthworm with special reference to the effects of cooling. *J. exp. Biol.* **21,** 46–57.

ANDERSON, J. C. (1956). Relations between metabolism and morphogenesis during regeneration in *Tubifex tubifex* II. *Biol. Bull. Wood's Hole,* **110,** 179–189.

ARBIT, J. (1957). Diurnal cycles and learning in earthworms. *Science,* **126,** 654–655.

AROS, B. and VÍGH, B. (1959). Changes in the neurosecretion of the nervous system of earthworm under various external conditions. *Acta. Biol. Hung.* Supp. 3. 47.

AROS, B. and BODNÁR, E. *Biol. Csop. Kozl.* 2. See TÖRÖ.

ARRHENIUS, O. (1921). Influence of soil reaction on earthworms. *Ecology.* **2,** 255–257.

AUCLAIR, J. L., HERLANT-MEEWIS, H. and DEMERS, M. (1951). Analyse qualitative des acides aminés de deux oligochaètes, *Aeolosoma hemprichi* et *Aeolosoma variegatum. Rev. Canad. Biol.* **10,** 162–166.

AVEL, M. (1929). Recherches expérimentales sur les caractères sexuels somatiques des lombriciens. *Bull. Biol.* **63**, 149–320.

BAHL, K. N. (1945). Studies on the structure, development and physiology of the nephridia of oligochaeta. VI. The physiology of excretion and the significance of the enteronephric type of nephridial system in Indian earthworms. *Quart. J. micr. Sci.* **85**, 343–389.

BAHL, K. N. (1947a). Studies on the structure, development and physiology of the nephridia of oligochaeta. VIII. Biochemical estimations of nutritive and excretory substances in the blood and coelomic fluid of the earthworm, and their bearing on the role of the two fluids in metabolism. *Quart. J. micr. Sci.* **87**, 357–371.

BAHL, K. N. (1947b). Excretion in the oligochaeta. *Biol. Revs.* **22**, 109–147.

BAHL, K. N. and LAL, M. B. (1933). On the occurrence of 'Hepatopancreatic' glands in the Indian earthworms of the genus *Eutyphoeus* Mich. *Quart. J. micr. Sci.* **76**, 107–127.

BAILEY, P. L. (1930). The influence of the nervous system in the regeneration of *Eisenia foetida* Savigny. *J. exp. Zool.* **57**, 473–509.

BALDWIN, E. (1949). *An Introduction to Comparative Biochemistry*, (third edition). pp. 164. Cambridge Univ. Press.

BALDWIN, F. M. (1917). Diurnal activity of the earthworm. *J. anim. Behav.* **7**, 187–190.

BEADLE, L. C. (1957). Respiration of the African swampworm *Alma emini* Mich. *J. exp. Biol.* **34**, 1–10.

BEATTY, I. M., MAGRATH, D. I. and ENNOR, A. H. (1959). Occurrence of D-serine in Lombricine. *Nature, Lond.* **183**, 591.

BEIDERMANN, W. (1904). Studien zur vergleichenden Physiologie der peristaltischen Bewegungen I. Die peristaltischen Bewegungen der Würmer und der Tonus glatter Muskeln. *Pflüg. Arch. ges. Physiol.* **102**, 475–542.

BERGMANN, W. (1949). Comparative biochemical studies on the lipids of marine invertebrates, with special reference to the sterols. *J. mar. Res.* **8**, 137–176.

BEVELANDER, G. (1952). Calcification in molluscs. III. Intake and deposition of $Ca^{45}$ and $P^{32}$ in relation to shell formation. *Biol. Bull. Wood's Hole*, **102**, 9–15.

BEVELANDER, G. and NAKAHARA, H. (1959). A histochemical and cytological study of the calciferous glands of *Lumbricus terrestris*. *Physiol. Zoöl.* **32**, 40–46.

BLASCHKO, H. and HIMMS, J. M. (1953). Amine oxidase in the earthworm. *J. Physiol.* **120**, 445–448.

BODENHEIMER, F. S. (1935). Soil conditions which limit earthworm distribution. *Zoogeografica.* **2**, 572–578.

BRADWAY, W. E. and MOORE, A. R. (1940). The locus of the action of the galvanic current in the earthworm, *Lumbricus terrestris*. *J. cell. comp. Physiol.* **15**, 47–54.

# REFERENCES

BRANDENBURG, J. (1956). Neurosekretorische Zellen des Regenwurms. *Naturwissenschaften*, **43**, 453.

BRETNALL, G. H. (1927). Earthworms and spectral colours. *Science*, **66**, 427.

VAN BRINK, J. M. and RIETSEMA, J. (1949). Some experiments on the active uptake of chlorine ions by the earthworm (*Lumbricus terrestris* L.). *Physiol. comp. et oecol.* **1**, 348–351.

BRINKMAN, R. and JONXIS, J. H. P. (1937). Alkali resistance and spreading velocity of foetal and adult types of mammalian haemoglobin. *J. Physiol.* **88**, 162–166.

BROWN, F. A. Jr. (1954). Simple, automatic, continuous-recording respirometer. *Rev. Sci. Instrum.* **25**, 415–417.

BULLOCK, T. H. (1945). Functional organization of the giant fibre system of *Lumbricus*. *J. Neurophysiol.* **8**, 55–71.

BULLOCK, T. H. (1947). Problems in invertebrate electrophysiology. *Physiol. Revs.* **27**, 643–664.

BULLOCK, T. H. (1951). Facilitation of conduction rate in nerve fibres. *J. Physiol.* **114**, 89–97.

BULLOCK, T. H. and TURNER, R. S. (1950). Events associated with conduction failure in nerve fibres. *J. cell. comp. Physiol.* **36**, 59–81.

BULLOCK, T. H., COHEN, M. J. and FAULSTICK, D. (1950). Effect of stretch on conduction in single nerve fibres. *Biol. Bull. Wood's Hole*, **99**, 320.

CARTER, G. S. (1940). *A General Zoology of the Invertebrates.* 4th ed. pp. 421. Sidgwick & Jackson.

CHAPMAN, G. (1958). The hydrostatic skeleton in the invertebrates. *Biol. Revs.* **33**, 338–371.

CLARK, A. M. (1957). The distribution of carbonic anhydrase in the earthworm and snail. *Aust. J. Sci.* **19**, 205–207.

COHEN, S. and LEWIS, H. B. (1949a). Nitrogenous metabolism of the earthworm (*Lumbricus terrestris*). *Fed. Proc.* **8**, 191.

COHEN, S. and LEWIS, H. B. (1949b). The nitrogenous metabolism of the earthworm (*Lumbricus terrestris*). *J. biol. Chem.* **180**, 79–91.

COHEN, S. and LEWIS, H. B. (1950). The nitrogenous metabolism of the earthworm (*Lumbricus terrestris*). *J. biol. Chem.* **184**, 479–484.

COLE, L. C. (1957). Biological clock in the unicorn. *Science*, **125**, 874–876.

COLLIER, J. C. (1947). Relations between metabolism and morphogenesis during regeneration in *Tubifex tubifex*. *Biol. Bull. Wood's Hole*, **92**, 167–177.

COLLIER, H. O. J. (1938). The immobilization of locomotory movements in earthworm, *Lumbricus terrestris*. *J. exp. Biol.* **15**, 339–357.

COLLIER, H. O. J. (1939a). Central nervous activity in the earthworm. 1. Responses to tension and to tactile stimulation. *J. exp. Biol.* **16**, 286–299.

COLLIER, H. O. J. (1939b). Central nervous activity in the earthworm. 2. Properties of the tension reflex. *J. exp. Biol.* **16**, 300–312.

COMBAULT, A. (1907). Quelques expériences pour déterminer le rôle des glandes calcifères des Lombrics. *C. R. Soc. Biol., Paris*, **62,** 440–442.

COMBAULT, A. (1909). Contribution à l'étude de la respiration et de la circulation de lombriciens. *J. Anat., Paris*, **45,** 358–399.

CONROY, D. A. (1960). Preliminary studies into the effect of temperature on the contraction rate of the dorsal blood vessels in the earthworm. *Bol. R. Soc. Española Hist. Nat. Biol.* **58,** 75–78.

CORDIER, R. (1934). Études histophysiologiques sur la néphridie du Lombric. *Archiv. Biol.* **45,** 431–471.

CUÉNOT, L. (1898). Études physiologiques sur les oligochètes. *Arch. Biol., Paris*, **15,** 79–124.

DANIELLI, J. F. and PANTIN, C. F. A. (1950). Alkaline phosphatase in protonephridia of terrestrial nemertines and planarians. *Quart. J. Mic. Sci.* **91,** 209–214.

DARWIN, C. R. (1881). The formation of vegetable mould through the action of worms with observations on their habits. John Murray & Co., London.

DAUSEND (1931). Über die Atmung der Tubificiden. *Z. vergl. Physiol.* **14,** 557–608.

DAVIS, J. G. and SLATER, W. K. (1928). The anaerobic metabolism of the earthworm (*Lumbricus terrestris*). *Biochem. J.* **22,** 338–345.

DAWSON, A. B. (1920). The intermuscular nerve cells of the earthworm. *J. comp. Neurol.* **32,** 155–171.

DELAUNAY, H. (1934). Le métabolisme de l'ammoniaque d'après les recherches relatives aux invertébrés. *Ann. Physiol. Physicochim. Biol.* **10,** 695–725.

DENNELL, R. (1949). Earthworm chaetae. *Nature, Lond.* **164,** 370.

DHÉRÉ, C. (1932). Sur la porphyrine tégumentaire du *Lumbricus terrestris*. *C. R. Acad. Sci., Paris*, **195,** 1436–1438.

DOBSON, R. M. and SATCHELL, J. E. (1956). *Eophila oculata* at Verulamium: a Roman Earthworm Population? *Nature, Lond.* **177,** 796–797.

DOLK, H. E. and VAN DER PAAUW, F. (1929). Die Leistungen des Hämoglobins beim Regenwurm. *Z. vergl. Physiol.* **10,** 324–343.

DOTTERWEICH, H. (1933). Die Funktion tierischer Kalkablagerungen als Pufferreserve im Dienste der Reaktionsregulation. Die Kalkdrüsen des Regenwurms. *Pflüg. Arch. ges. Physiol.* **232,** 263–286.

DURCHON, M. and LAFON, M. (1951). Quelques données biochimiques sur les annélides. *Ann. Sci. Nat., Zool.* 11 ser. **13,** 427–452.

ECCLES, J. C., GRANIT, R. and YOUNG, J. Z. (1933). Impulses in the giant fibres of earthworms. *J. Physiol.* **77,** 23–25.

ENDERS, E. (1952). Die Wirkung körpereigener Organextrakte auf die Bewegungen des Regenwurmdarmes. *Ann. Univ. Sarav. Natur.* **1,** 294–308.

ENNOR, A. H. and MORRISON, J. F. (1958). Biochemistry of the phosphagens and related guanidines. *Physiol. Rev.* **38,** 631–674.

ENNOR, A. H., ROSENBERG, H., ROSSITER, R. J., BEATTY, I. M. and GAFFNEY, T. (1960). The isolation and characterization of D-serine ethanolamine phosphodiester from earthworms. *Biochem. J.* **75**, 179–182.

EULER, U. S. VON (1948). Preparation, purification and evaluation of noradrenaline, and adrenaline in organ extracts. *Arch. Int. Pharmacodyn Ther.* **77**, 477–485.

EULER, U. S. VON (1961). Occurrence of Catecholamines in Acrania and Invertebrates. *Nature, Lond.* **190**, 170–171.

EWER, D. W. and HANSON, J. (1945). Some staining reactions of invertebrate mucoproteins. *J. R. micr. Soc.* **65**, 40–43.

FLORKIN, M. (1957). Biochimie et evolution animale, *Verh. schweiz. naturf. Ges.* 35–53.

FLORKIN, M. and DUCHATEAU, G. (1943). Les formes du système enzymatique de l'uricolyse et l'évolution du catabolisme purique chez les animaux. *Arch. int. Physiol.* **53**, 267–307.

FOX, H. M. (1940). Function of chlorocruorin in *Sabella* and of haemoglobin in *Lumbricus*. *Nature, Lond.* **145**, 781–782.

FOX, H. M. (1945). The oxygen affinities of certain invertebrate haemoglobins. *J. exp. Biol.* **21**, 161–165.

FOX, H. M. and TAYLOR, A. E. R. (1955). The tolerance of oxygen by aquatic invertebrates. *Proc. roy. Soc.* B. **143**, 214–225.

FRENCH, C. E., LISCINSKY, S. A. and MILLER, D. R. (1957). Nutrient composition of earthworms. *J. Wildlife Mgmt.* **21**, 348.

FRIEDLÄNDER, B. (1894). Beiträge zur Physiologie des Centralnervensystems und des Bewegungsmechanismus der Regenwürmer. *Pflüg. Arch. ges. Physiol.* **58**, 168–206.

GABBAY, K. H. (1958). An investigation of the calciferous glands in *Lumbricus terrestris*. *Biol. Rev. City Coll. New York*, **21**, 16–19.

VAN GANSEN P. SEMAL (1956). Les cellules chloragogènes des Lombriciens. *Bull. Biol.* **90**, 335–356.

VAN GANSEN, P. SEMAL (1957a). Histophysiologie du tube digestif d'*Eisenia foetida* Sav. region buccale, pharynx et glandes pharyngiennes. *Bull. Biol.* **91**, 225–239.

VAN GANSEN, P. SEMAL (1957b). Le lipopigment des chloragosomes des lombriciens. *Ann. d'histochimie*, **2**, 41–55.

VAN GANSEN, P. SEMAL (1958). Physiologie des cellules chloragogènes d'un lombricien. *Enzymologia*, **20**, 98–108.

VAN GANSEN, P. SEMAL (1959a). Structure des glandes calciques d'*Eisenia foetida* Sav. *Bull. Biol.* **93**, 38–63.

VAN GANSEN, P. SEMAL (1959b). Structure du jabot et du gesier d'*Eisenia foetida* (Lombricien). *Ann. Soc. Roy. zool. Belg.* **89**, 341–363.

VAN GANSEN, P. SEMAL (1960). Occurrence of a non-fibrillar elastin in the earthworm. *Nature, Lond.* **186**, 654–655.

VAN GANSEN, P. SEMAL and VAN DER MEERSCHE, G. (1958). L'ultrastructure des cellules chloragogènes. *Bull. Mic. Appliq.* **8**, 7–13.

VAN GANSEN, P. SEMAL, VAN DER MEERSCHE, G. and CASTIAUX, P. (1959). L'ultrastructure des glandes calciques d'*Eisenia foetida* sav. *Bull. Mic. Appliq.* **9,** 33–37.

GARREY, W. L. and MOORE, A. R. (1915). Peristalsis and co-ordination in the earthworm. *Amer. J. Physiol.* **39,** 139–148.

GASKELL, J. F. (1914). Chromaffin system of annelids and reaction of this system to the contractile vascular system in the leach *Hirudo medicinalis. Philos. Trans.* B, **205,** 153–212.

GASKELL, J. F. (1920). Adrenalin in annelids. *J. gen. Physiol.* **2,** 13–85.

GODEAUX, J. (1954). Recherches électrophorétiques sur les protéines musculaires du lombric. *Bull. Acad. Belg. Cl. Sci.* **40,** 948–961.

GOODRICH, E. S. (1945). The study of nephridia and genital ducts since 1895. *Quart. J. micr. Sci.* **86,** 113–392.

GRAY, J. and LISSMANN, H. W. (1938). Studies in animal locomotion. VII. Locomotory reflexes in the earthworm. *J. exp. Biol.* **15,** 506–517.

GUARDABASSI, A. (1957). La ghiandole calcifer di *Eisenia foetida* studio isto et citochimico. *Z. Zellforsch.* **46,** 619–634.

HAGGAG, G. and KHALAF EL-DUWEINI, A. (1959). Main nitrogenous constituents of the excreta and tissues of earthworms. *Proc. Egypt. Acad. Sci.* **13,** 1–5.

HAMA, K. (1959). Some observations on the fine structure of the giant nerve fibres of the earthworm *Eisenia foetida*. *J. Biophys. Biochem. Cytol.* **6,** 61–66.

HANSON, J. (1957). The structure of the smooth muscle fibres in the body wall of the earthworm. *J. Biophys. Biochem. Cytol.* **3,** 111–122.

HANSON, J. and LOWY, J. (1960). Contractile apparatus in invertebrate animals. In *Structure and Function of Muscle* (Ed. G. Bourne, Academic Press, New York).

HANSTROM, B. (1928). *Vergleichende Anatomie des Nervensystems der wirbellosen Tiere*. Springer-Verlag, Berlin.

HARKER, J. E. (1958). Diurnal rhythms in the animal kingdom. *Biol. Revs.* **33,** 1–52.

HARKER, J. E. (1960). Internal factors controlling the sub-oesophageal ganglion neurosecretory cycle in *Periplaneta americana* L. *J. exp. Biol.* **37,** 164–170.

HARMS, W. R. (1948). Über ein inkretorisches Cerebralorgan bei Lumbriciden sowie Beschreibung eines verwandten Organs bei drei neuen *Lycastis* Arten. *Arch. EntwMech. Org.* **143,** 332–346.

HARRINGTON, N. R. (1899). The calciferous glands of the earthworm, with appendix on the circulation. *J. Morph.* **15,** suppl.

HAUGHTON, T. M., KERKUT, G. A. and MUNDAY, K. A. (1958). The oxygen dissociation and alkaline denaturation of haemoglobins from two species of earthworm. *J. exp. Biol.* **35,** 360–368.

HAWKE, P. B., OSER, B. L. and SUMMERSON, W. H. (1954). *Practical Physiological Chemistry*. pp. 1439. Churchill, London.

HECK, L. (1920). Über die Bildung einer Assoziation beim Regenwurm auf Grund von Dressurversuchen. *Lotos*, **67–8,** 169–189.

# REFERENCES

HEIDERMANNS, C. (1937). Über die Harnstoffbildung beim Regenwurm. *Zool. Jahrb. Allgem. Zool. Physiol. Tiere*, **58**, 57–68.

HERLANT-MEEWIS, H. (1953). Contribution à l'étude de la régénération chez les oligochetes aéolosomatidae. *Ann. Soc. Roy. Zool. Belg.* **84**, 117–159.

HERLANT-MEEWIS, H. (1954). Croissance et reproduction du lombricien *Eisenia foetida*. *Ann. Soc. Roy. Zool. Belg.* **85**, 119–151.

HERLANT-MEEWIS, H. (1956). Croissances et neurosécrétion chez *Eisenia foetida* (Sav.). *Ann. Sci. Nat. Zool. et Biol. Animale.* 11th ser. **18**, 185–198.

HERLANT-MEEWIS, H. (1957). Reproduction et neurosécrétion chez *Eisenia foetida* (Sav.) *Ann. Soc. Roy. Zool. Belg.* **87**, 151–185.

HESS, W. N. (1924). Reactions to light in the earthworm, *Lumbricus terrestris*. *J. Morph.* **39**, 515–542.

HESS, W. N. (1925a). Nervous system of the earthworm, *Lumbricus terrestris*. *J. Morph.* **40**, 235–260.

HESS, W. N. (1925b). Photoreceptors of *Lumbricus terrestris*, with special reference to their distribution, structure and function. *J. Morph.* **41**, 63–93.

HODGSON, E. S. (1955). Problems in invertebrate chemoreception. *Quart. Rev. Biol.* **30**, 331–347.

HOGBEN, L. and KIRK, R. L. (1944). Studies on temperature regulation. 1. The pulmonata and oligochaeta. *Proc. roy. Soc. B*, **132**, 239–252.

HORRIDGE, G. A. and ROBERTS, M. B. V. (1960). Neuro-muscular transmission in the earthworm. *Nature, Lond.* **186**, 650.

HOWELL, C. D. (1939). The responses to light in the earthworm, *Pheretima agrestis* Goto and Hatai, with special reference to the function of the nervous system. *J. exp. Zoöl.* **81**, 231–259.

HUBL, H. (1953). Die inkretorischen Zellelemente im Gehirn der *lumbriciden*. *Arch. EntwMech. Org.* **146**, 421–432.

HUBL, H. (1956). Über die Beziehungen der Neurosekretion zum Regenerationsgeschehen bei *lumbriciden* nebst Beschreibung eines neuartigen neurosekretorischen Zelltyps im Unterschlundganglion. *Arch. EntwMech. Org.* **149**, 73–87.

HURWITZ, S. N. (1910). The reactions of earthworms to acids. *Proc. Amer. Acad.* **46**, 67–81.

HYMAN, L. H. (1916). An analysis of the process of regeneration in certain microdrilous oligochaetes. *J. exp. Zool.* **20**, 99–163.

ISSIDORIDES, M. (1956). Ultrastructure of the synapse in the giant axons of the earthworm. *Exp. Cell Res.* **11**, 423–436.

JACKSON, C. M. (1926). Storage of water in various parts of the earthworm at different stages of exsiccation. *Proc. Soc. exp. Biol., N.Y.* **23**, 500–504.

JEFFERSON, P. (1956). Studies on the earthworms of turf. *J. Sports Turf Research Inst.* **9**, 166–179.

JEWELL, M. E. and LEWIS, H. B. (1918). The occurrence of lichenase in the digestive tract of invertebrates. *J. biol. Chem.* **33**, 161–167.

JOHNSON, M. L. (1942). The respiratory function of the haemoglobin of the earthworm. *J. exp. Biol.* **18,** 266–277.

JORDAN, H. and SCHWARZ, B. (1920). Einfache Apparate zur Gasanalyse und Mikrorespirometrie in bestimmten Gasgemischen, und über die Bedeutung des Hämoglobins beim Regenwurm. *Pflüg. Arch. ges. Physiol.* **185,** 311–321.

KAGAWA, KEN-ITI (1949). The distribution of pH in the alimentary tract of earthworms. *Tôhoku Imp. Univ. Sci. Reps.* 4th Ser. **18,** 163–165.

KALMUS, H., SATCHELL, J. E. and BOWEN, J. C. (1955). On the colour forms of *Allolobophora chlorotica* Sav. (Lumbricidae). *Ann. Mag. Nat. Hist.* Ser. 12, **8,** 795–800.

KAMAT, D. N. (1955). The nature of the intestinal proteinase of the earthworm, *Pheretima elongata* (Steph). *J. An. Morph. Physiol.* **2,** 79–86.

KAO, C. Y. (1956). Basis for after-discharge in the median giant axon of the earthworm. *Science,* **123,** 803.

KAO, C. Y. and GRUNDFEST, H. (1957). Postsynaptic electrogenesis in septate giant axons. 1. Earthworm median giant axon. *J. Neurophysiol.* **20,** 553–573.

KEILIN, D. (1920). On the pharyngeal or salivary gland of the earthworm. *Quart. J. micr. Sci.* **65,** 33–61.

KENNEDY, G. Y. (1959). A porphyrin pigment in the integument of *Arion ater* (L). *J. mar. biol. Ass. U.K.* **38,** 27–32.

KENNEDY, G. Y. (1956). Porphyrin pigments in the tectibranch mollusc *Akera bullata. J. mar. biol. Ass. U.K.* **35,** 35–40.

KENNEDY, G. Y. and VEVERS, H. G. (1953). The biology of *Asterias rubens.* V. A porphyrin pigment in the integument. *J. mar. biol. Ass. U.K.* **32,** 235–247.

KERKUT, G. A. (1960). *Implications of Evolution.* pp. 174. Pergamon Press, Oxford and London.

KERKUT, G. A. and TAYLOR, B. J. R. (1956). The sensitivity of the pedal ganglion of the slug to osmotic pressure changes. *J. exp. Biol.* **33,** 493–501.

KIRBERGER, C. (1953). Untersuchungen über die Temperaturabhängigkeit von Lebensprozessen bei verschiedenen Wirbellosen. *Z. vergl. Physiol.* **35,** 175–198.

KISS, I. (1957). The invertase activity of earthworm casts and soils from ant hills. *Agrokémia és Talajtan,* **6,** 65–68.

KOBAYASHI, S. (1928). The spectroscopic observations on porphyrin found in the integument of earthworm, *Allolobophora foetida* (Sav.) *Tôhoku Imp. Univ. Sci. Reps.* 4th Ser. **3,** 467–480.

KOMINZ, D. R., SAAD, F. and LAKI, K. (1957). Vertebrate and invertebrate tropomyosins. *Nature, Lond.* **179,** 206.

KOPENHAVER, M. E. (1937). Axial differences in water-absorbing properties of oligochaete tissues. *Physiol. Zoöl.* **10,** 315–326.

KRÜGER, F. (1952). Über die Beziehung des Sauerstoffverbrauchs zum Gewicht bei *Eisenia foetida* (Sav.). *Z. vergl. Physiol.* **34,** 1–5.

# REFERENCES

KRÜGER, F. and BECKER, H. (1940). Beiträge zur Physiologie des Hämoglobins wirbelloser Tiere. II. Vergleich der Atmung normaler und mit Kohlenoxyd vergifteter Regenwürmer bei einer Temperatur von 9 °C. *Z. vergl. Physiol.* **28**, 180–196.

KUENZER, P. (1958). Verhaltensphysiologische Untersuchungen über das Zucken des Regenwurms. *Z. Tierpsychol.* **15**, 31–49.

KURTZ, I. and SCHRANK, A. R. (1955). Bioelectrical properties of intact and regenerating earthworms, *Eisenia foetida*. *Physiol. Zoöl.* **28**, 322–330.

LANGDON, F. E. (1895). The sense organs of *Lumbricus agricola* Hoffm. *J. Morph.* **11**, 193–234.

LAVERACK, M. S. (1960a). The identity of the porphyrin pigments of the integument of earthworms. *Comp. Biochem. Physiol.* **1**, 259–266.

LAVERACK, M. S. (1960b). Tactile and chemical perception in earthworms. I. Responses to touch, sodium chloride, quinine and sugars. *Comp. Biochem. Physiol.* **1**, 155–163.

LAVERACK, M. S. (1961a). II. Responses to acid pH solutions. *Comp. Biochem. Physiol.* **2**, 22–34.

LAVERACK, M. S. (1961b). Effect of temperature changes on the spontaneous nervous activity of the isolated nerve cord of *Lumbricus terrestris*. *Comp. Biochem. Physiol.* **3**, 136–140.

LAWRENCE, R. D. and MILLAR, H. R. (1945). Protein content of earthworms. *Nature, Lond.* **155**, 517.

LESSER, E. J. (1908). Chemische Prozesse bei Regenwürmern. 1. Der Hungerstoffwechsel. *Z. Biol.* **50**, 421–425.

LESSER, E. J. and TASCHENBERG, E. W. (1908). Über Fermente des Regenwurms. *Z. Biol.* **50**, 446–470.

DE LEY, J. and VERCRUYSSE, R. (1955). Glucose-6-phosphate and gluconate-6-phosphate dehydrogenase in worms. *Biochim. Biophys. Acta.* **16**, 615–616.

LIEBMANN, E. (1942a). The role of the chloragogue in regeneration of *Eisenia foetida* (Sav.). *J. Morph.* **70**, 151–187.

LIEBMANN, E. (1942b). The coelomocytes of lumbricidae. *J. Morph.* **71**, 221–249.

LIEBMANN, E. (1946). On trephocytes and trephocytosis: a study on the role of leucocytes in nutrition and growth. *Growth.* **10**, 291–330.

MACMUNN, C. A. (1886). On the presence of haematoporphyrin in the integument of certain invertebrates. *J. Physiol.* **7**, 240–252.

MÁLEK, R. (1927). Associatives Gedächtnis bei den Regenwürmern. *Biol. gen.* **3**, 317–328.

MALOEUF, N. S. R. (1936a). The neurogenic basis of "crawling". *Z. vergl. Physiol.* **56**, 379–380.

MALOEUF, N. S. R. (1936b). The "Metabolic gradient" in the adult earthworm. *Z. vergl. Physiol.* **56**, 429–436.

MALOEUF, N. S. R. (1939). The volume and osmo-regulative functions of the alimentary tract of the earthworm (*Lumbricus terrestris*) and on the absorption of chloride from freshwater by this animal. *Zool. Jahrb. Abt. Allgem. Zool. Physiol. Tiere*, **59**, 544–552.

MALOEUF, N. S. R. (1940a). The volume and osmo-regulative functions of the alimentary tract of the earthworm (*Lumbricus terrestris*) and on the absorption of chloride from freshwater by this animal. *Z. vergl. Physiol.* **59**, 535–552.

MALOEUF, N. S. R. (1940b). Osmo- and volume regulation in the earthworm with special reference to the alimentary tract. *J. cell. comp. Physiol.* **16**, 175–187.

MANGOLD, O. (1951). Experimente zur Analyse des chemischen Sinns des Regenwurms. I. Methode und Verhalten zu Blättern von Pflanzen. *Zool. Jahrb. Abt. Allgem. Zool. Physiol. Tiere*, **62**, 441–512.

MANGOLD, O. (1953). Experimente zur Analyse des chemischen Sinns des Regenwurms. II. Versuche mit Chinin, Säuren und Sübstoffen. *Zool. Jahrb. Abt. Allgem. Zool. Physiol. Tiere*, **63**, 501–557.

MANWELL, C. (1959). Alkaline denaturation and oxygen equilibrium of annelid haemoglobins. *J. cell. comp. Physiol.* **53**, 61–74.

MARAPAO, B. P. (1959). The effect of nervous tissue extracts on neurosecretion in the earthworm, *Lumbricus terrestris. Catholic U. Amer., Biol. Stud.* **55**, 1–34.

MARTIN, A. W. (1957). Recent advances in knowledge of invertebrate renal function. *Invertebrate Physiol.* (Ed. B. T. Scheer) Univ. of Oregon Pub. pp. 247–276.

MARUYAMA, K. and KOMINZ, D. R. (1959). Earthworm myosin. *Z. vergl. Physiol.* **42**, 17–19.

MASSARO, E. J. and SCHRANK, A. R. (1959). Chemical inhibition of segment regeneration in *Eisenia foetida. Physiol. Zoöl.* **32**, 185–196.

MAST, S. O. (1917). The relation between spectral colours and stimulation in the lower organisms. *J. Exp. Zool.* **22**, 471–528.

M'DOWALL, J. (1926). Preliminary work towards a morphological and physiological study of the calciferous glands of the earthworm. *Proc. Phys. Soc. Edinb.* **21**, 65–72.

MELDRUM, N. U. and ROUGHTON, F. J. W. (1934). Carbonic anhydrase, its preparation and properties. *J. Physiol.* **80**, 113–170.

MENDES, E. G. and NONATO, E. F. (1957). The respiratory metabolism of tropical earthworms. II. Studies on cutaneous respiration. *Univ. Sao Paulo, Fac. filos. of ciênc e letras, Bol. Zool.* **21**, 153–166.

MENDES, E. G. and VALENTE, D. (1953). The respiratory metabolism of tropical earthworms. I. The respiratory rate and the action of carbon monoxide at normal oxygen pressure. *Bol. Univ. Sao Paulo*, **18**, 91–102.

MERKER, E. and BRÄUNIG, G. (1927). Die Empfindlichkeit feuchthäutiger Tiere im Lichte. III. Die Atemnot feuchthäutiger Tiere im Licht der Quarzquecksilberlampe. *Zool. Jahrb. Abt. Allgem. Zool. Physiol. Tiere*, **43**, 275–338.

MICHAELSON, W. (1895). Oligochaetan. *Kukenthal und Krumbach Handbuch*, **8**, 32.

MICHON, J. (1949). Influence de la dessiccation sur la diapause des lombriciens. *C. R. Acad. Sci., Paris*, **228**, 1455–1456.

MICHON, J. (1954). Influence de l'isolement à partir de la maturité sexuelle sur la biologie des lumbricidae. *C. R. Acad. Sci., Paris*, **238**, 2457–2458.

MICHON, J. (1957). Contribution expérimentale à l'étude de la biologie des lumbricidae. Les variations pondérales au cours des différentes modalités du développement postembryonnaire. *Année Biol.* **33**, 367–376.

MILLARD, A. and RUDALL, K. M. (1960). Light and electron microscope studies of fibres. *J. Roy. Mic. Soc.* **79**, 227–231.

MILLER, J. A. and TING, H. P. (1949). The role of the subepidermal nervous system in the locomotion of the earthworm. *Ohio J. Sci.* **49**, 109–114.

MILLOTT, N. (1943a). The visceral nervous system of the earthworm. I. Nerves controlling the tone of the alimentary canal. *Proc. roy. Soc.* B. **131**, 271–295.

MILLOTT, N. (1943b). The visceral nervous system of the earthworm. II. Evidence of chemical transmission and the action of sympathomimetic and parasympathomimetic drugs on the tone of the alimentary canal. *Proc. roy. Soc.* B. **131**, 362–373.

MILLOTT, N. (1944). The visceral nerves of the earthworm. III. Nerves controlling secretion of protease in the anterior intestine. *Proc. roy. Soc.* B. **132**, 200–212.

MOMENT, G. B. (1949). On the relation between growth in length, the formation of new segments, and electric potential in an earthworm. *J. exp. Zool.* **112**, 1–12.

MOMENT, G. B. (1953). On the way a common earthworm, *Eisenia foetida*, grows in length. *J. Morph.* **93**, 489–507.

MOORE, A. R. (1921). Chemical stimulation of the nerve cord of *Lumbricus terrestris*. *J. gen. Physiol.* **4**, 29–31.

MOORE, A. R. (1923a). Muscle tension and reflexes in the earthworm. *J. gen. Physiol.* **5**, 327–333.

MOORE, A. R. (1923b). Galvanotropism in the earthworm. *J. gen. Physiol.* **5**, 453–459.

MORGAN, T. H. and DIMON, A. C. (1904). An examination of the problems of physiological polarity and electrical polarity in the earthworm. *J. exp. Zool.* **1**, 331–347.

MYOT, C. (1957). Étude de la glande de Morren chez deux oligochètes lombricides. *Arch. Zool. Exptle. Gén.* **94**, 61–87.

NEEDHAM, A. E. (1957). Components of nitrogenous excreta in the earthworms *Lumbricus terrestris* L. and *Eisenia foetida* (Savigny). *J. exp. Biol.* **34**, 425–446.

NEEDHAM, A. E. (1958). The pattern of nitrogen-excretion during regeneration in Oligochaetes. *J. exp. Zool.* **138**, 369–430.

NEEDHAM, A. E. (1960). The arginase activity of the tissues of the earthworms *Lumbricus terrestris* L., and *Eisenia foetida* (Savigny). *J. exp. Biol.* **37,** 775–782.

NEEDHAM, A. E. (1962). Arginase activity in earthworms. *Comp. Biochem. Physiol.* **5,** 96–103.

NEWELL, G. E. (1950). The role of the coelomic fluid in the movements of earthworms. *J. exp. Biol.* **27,** 110–121.

NICOL, J. A. C. (1948). The giant axons of annelids. *Quart. Rev. Biol.* **23,** 291–323.

NOMURA, E. (1926). Effect of light on the movements of the earthworm *Allolobophora foetida* (Sav.). *Tôhoku Imp. Univ. Sci. Rep.* 4th Ser. **1.** 293–409.

O'BRIEN, B. R. A. (1947). Studies in the metabolism of normal and regenerating tissue of the earthworm. Part 1. Factors affecting the endogenous oxygen consumption of normal and regenerating muscle tissue. *Proc. Linn. Soc. N.S.W.* **72,** 367–378.

O'BRIEN, B. R. A. (1957a). Evidence in support of an axial metabolic gradient in the earthworm. *Aust. J. exp. Biol. med. Sci.* **35,** 83–92.

O'BRIEN, B. R. A. (1957b). Tissue metabolism during posterior regeneration in the earthworm. *Aust. J. exp. Biol. med. Sci.* **35,** 373–380.

OGAWA, F. (1928). On the number of ganglion cells and nerve fibres in some of the ventral nerve cords of the earthworm. *Tôhoku Imp. Univ. Sci. Rep.* 4th Ser. **3,** 745–756.

OSTLUND, E. (1954). The distribution of catechol amines in lower animals and their effect on the heart. *Acta physiol. scand.* **31,** Supp. 112, p. 67.

PANT, R. (1959). Isolation of lombricine and its enzymic phosphorylation. *Biochem. J.* **73,** 30–33.

PANTIN, C. F. A. (1947). The nephridia of *Geonemertes dendyi*. *Quart. J. micr. Sci.* **88,** 15–25.

PARKER, G. H. and PARSHLEY, H. M. (1911). The reactions of earthworms to dry and to moist surfaces. *J. exp. Zool.* **11,** 361–363.

PARLE, J. N. (1960). Personal communication.

PERKINS, M. (1929). Growth gradients and the axial relations of the animal body. *Nature, Lond.* **124,** 299–300.

PESCHEN, K. E. (1939). Untersuchungen über das Vorkommen und den Stoffwechsel des Guanins im Tierreich. *Zool. Jber.* **59,** 429–462.

PETRUCCI, D. (1952). Prime osservazioni sul metabolismo dei chetoacidi negli oligocheti. *Boll. Zool.* **19,** 145–155.

PETRUCCI, D. (1954). Richerche sul ciclo degli acidi tricarbossilici negli oligocheti. *Riv. Biol.* **46,** 241–251.

PETRUCCI, D. (1955). Citocromossidasi e respirazione in presenza di KCN e di CO negli Oligocheti. *Monit. Zool. ital.* **63,** 143–163.

PICKEN, L. E. R., PRYOR, M. G. M. and SWANN, M. M. (1947). Orientation of fibrils in natural membranes. *Nature, Lond.* **159,** 434.

POMERAT, C. M. and ZARROW, M. X. (1936). The effect of temperature on the respiration of the earthworm. *Proc. Nat. Acad. Sci., Wash.* **22,** 270–273.

# REFERENCES

Powell, V. E. (1951). Alkaline phosphatase in the regenerating annelid. *Anat. Rec.* **111**, 101–107.

Prosser, C. L. (1933). Correlation between development of behaviour and neuromuscular differentiation in embryos of *Eisenia foetida* Sav. *J. comp. Neurol.* **58**, 603–641.

Prosser, C. L. (1934a). The nervous system of the earthworm. *Quart. Rev. Biol.* **9**, 181–200.

Prosser, C. L. (1934b). Effect of the central nervous system on responses to light in *Eisenia foetida* Sav. *J. comp. Neurol.* **59**, 61–92.

Prosser, C. L. (1935). Impulses in the segmental nerves of the earthworm. *J. exp. Biol.* **12**, 95–104.

Prosser, C. L., Brown, F. A., Bishop, D. W., Jahn, T. L. and Wulff, V. J. (1950). *Comparative Animal Physiology.* 888 pp. Saunders, New York.

Prosser, C. L. and Zimmerman, G. L. (1943). Effects of drugs on the hearts of *Arenicola* and *Lumbricus. Physiol. Zoöl.* **16**, 77–83.

Przlecki, S. J. (1923). L'excrétion ammoniacale sur les invertebrés dans les conditions normales et expérimentales. *Arch. Int. Physiol.* **20**, 103–110.

Puytorac, P. De and Mauret, P. (1956). Détermination de certaines des conditions écologiques propres aux différents ciliés parasites du tube digestif d'*Allolobophora savigny* G. et H. (Oligochète). *Bull. Biol.* **90**, 123–141.

Raffy, A. (1930). La respiration des vers de terre dans l'eau. Action de la teneur en oxygène et de la lumiére sur l'intensité de la respiration pendant l'immersion. *C. R. Soc. Biol., Paris*, **105**, 862–864.

Ralph, C. L. (1957). Persistent rhythms of activity and $O_2$ consumption in the earthworm. *Physiol. Zoöl.* **30**, 41–44.

Ramsay, J. A. (1949a). The osmotic relations of the earthworm. *J. exp. Biol.* **26**, 46–56.

Ramsay, J. A. (1949b). The site of formation of hypotonic urine in the nephridium of *Lumbricus. J. exp. Biol.* **26**, 65–75.

Ratner, S. C. and Miller, K. R. (1959). Classical conditioning in earthworms, *Lumbricus terrestris. J. comp. Physiol. Psychol.* **52**, 102–105.

Reed, R. and Rudall, K. M. (1948). Electron microscope studies on the structure of earthworm cuticles. *Biochim. Biophys. Acta.* **2**, 7–18.

Rey, C. (1956). Les esters phosphorés des muscles du lombric. *Biochim. Biophys. Acta.* **19**, 300–307.

de Robertis, E. D. P. and Bennett, H. S. (1955a). Some features of fine structure of cytoplasm of cells in the earthworm nerve cord. *Union Int. Sci. Biol. B.* **21**, 261–273.

de Robertis, E. D. P. and Bennett, H. S. (1955b). Some features of the sub-microscopic morphology of synapses in frog and earthworm. *J. Biophys. Biochem. Cytol.* **1**, 47–58.

Roberts, M. B. V. (1960). Giant fibre reflex of the earthworm. *Nature, Lond.*, **186**, 167.

ROBERTSON, J. D. (1936). The function of the calciferous glands of earthworms. *J. exp. Biol.* **13**, 279–297.

ROBINET, C. (1883). Recherches physiologique sur la sécrétion des glandes de Morren du *Lumbricus terrestris*. *C. R. Acad. Sci., Paris*, **97**, 192–194.

ROGERS, C. G. and LEWIS, E. M. (1914). The relation of the body temperature of the earthworm to that of its environment. *Biol. Bull. Wood's Hole*, **27**, 262–268.

ROOTS, B. I. (1955). The water relations of earthworms. 1. The activity of the nephridiostome cilia of *Lumbricus terrestris* L. and *Allolobophora chlorotica* Savigny, in relation to the concentration of the bathing medium. *J. exp. Biol.* **32**, 765–774.

ROOTS, B. I. (1956). The water relations of earthworms. 2. Resistance to desiccation, immersion and behaviour when submerged and when allowed a choice of environment. *J. exp. Biol.* **33**, 29–44.

ROOTS, B. I. (1957). Nature of chloragogen granules. *Nature, Lond.* **179**, 679–680.

ROOTS, B. I. (1960). Some observations on the chloragogenous tissue of earthworms. *Comp. Biochem. Physiol.* **1**, 218–226.

ROOTS, B. I. and PHILLIPS, R. R. (1960). Burrowing and the action of the pharynx of earthworms. *Med. Biol. Illus.* **10**, 28–31.

ROŞCA, D. I., WITTENBERGER, C. and RUSDEA, D. (1958). Comportarea la variatii de salinitate. XLV. Cercetări asupra osmoregulării şi a rolului sistemului nervos in fenomenele de osmo-riegulare la *Hirudo medicinalis*. *Stud. Cercetaria Biol. (CLVJ.)* **9**, 113–136.

ROSENBERG, H. and ENNOR, A. H. (1959). The isolation of lombricine and its possible biological precursor. *Biochem. J.* **73**, 521–526.

ROSENBERG, H. and ENNOR, A. H. (1960). Occurrence of free D-serine in the earthworm. *Nature, Lond.* **187**, 617–618.

ROSSITER, R. J., GAFFNEY, T., ROSENBERG, H. and ENNOR, A. H. (1960). Biosynthesis of lombricine, *Nature, Lond.* **185**, 383–384.

ROSSITER, R. J., GAFFNEY, T. J., ROSENBERG, H. and ENNOR, A. H. (1960). The formation *in vivo* of lombricine in the earthworm *Megascolides cameroni*. *Biochem. J.* **76**, 603–610.

RUDALL, K. M. (1955). The distribution of collagen and chitin. *SEB. Symposium 9 "The fibrous proteins"*, 49–71.

RUSHTON, W. A. H. (1945). Action potentials from the isolated nerve cord of the earthworm. *Proc. roy. Soc.* B. **132**, 423–437.

RUSHTON, W. A. H. (1946). Reflex conduction in the giant fibres of the earthworm. *Proc. roy. Soc.* B. **133**, 109–120.

RUSHTON, W. A. H. and BARLOW, H. B. (1943). Single fibre response from an intact animal. *Nature, Lond.* **152**, 597–598.

RUSSELL, E. J. (1950). *Soil Conditions and Plant Growth*. (8th edition, Revised by E. W. Russell). Longmans, London.

SALOMON, K. (1941). Studies on invertebrate haemoglobins (Erythrocruorins). *J. gen. Physiol.* **24**, 367–375.

SAROJA, K. (1959). Studies on the oxygen consumption in tropical Poikilotherms. 2. Oxygen consumption in relation to body size and temperature in the earthworm *Megascolex mauritii* when kept submerged in water. *Proc. Ind. Acad. Sci.*, B. **49**, 183–193.

SATCHELL, J. E. (1956). Some aspects of earthworm ecology. *Soil Zoology*, 180–201.

SCHARRER, B. (1937). Über sekretorisch tätige Nervenzellen bei wirbellosen Tieren. *Naturwissenschaften*, **9**, 131–138.

SCHARRER, E. and SCHARRER, B. (1937). Über Drüsennervenzellen und neurosekretorische Organe bei Wirbellosen und Wirbeltieren. *Biol. Rev.* **12**, 185–216.

SCHEER, B. T. (1948). *Comparative Physiology*. 563 pp. Wiley, New York.

SCHMID, L. A. (1947). Induced neurosecretion in *Lumbricus terrestris*. *J. exp. Zool.* **104**, 365–377.

SCHMIDT, P. (1927). Anabiosis of the earthworm. *J. exp. Zool.* **27**, 57–72.

SCHNEIDERMAN, H. A. and GILBERT, L. I. (1958). Substances with juvenile hormone activity in crustacea and other invertebrates. *Biol. Bull. Wood's Hole*, **115**, 530–535.

SHEARER, C. (1924). On the oxygen consumption rate of parts of the chick embryo and fragments of the earthworm. *Proc. roy. Soc.* B. **96**, 146–156.

SHEARER, C. (1930). A re-investigation of metabolic gradients. *J. exp. Biol.* **7**, 260–268.

SHIRAISHI, K. (1954). On the chemotaxis of the earthworm to carbon dioxide. *Tôhoku Imp. Univ. Sci. Rep.*, 4th Ser. **20**, 356–361.

SINGLETON, L. (1957). The chemical structure of the earthworm cuticle. *Biochim. Biophys. Acta.* **24**, 67–72.

SJÖSTRAND, F. S. and RHODIN, J. (1953). The ultrastructure of the proximal convoluted tubules of the mouse kidney as revealed by high resolution electron microscopy. *Exp. Cell. Res.* **5**, 426–456.

SMALLWOOD, W. M. (1926). The peripheral nervous system of the common earthworm, *Lumbricus terrestris*. *J. comp. Neurol.*, **42**, 35–55.

SMALLWOOD, W. M. (1930). The nervous structure of the annelid ganglion. *J. comp. Neurol.* **51**, 377–392.

SMALLWOOD, W. M. and HOLMES, M. T. (1927). The neurofibrillar structure of the giant fibres in *Lumbricus terrestris* and *Eisenia foetida*. *J. comp. Neurol.* **43**, 327–345.

SMITH, A. C. (1902). The influence of temperature, odors, light and contact on the movements of the earthworm. *Amer. J. Physiol.* **6**, 459–486.

STEPHENSON, J. (1930). *The Oligochaeta*. pp. 978. Oxford Univ. Press.

STEPHENSON, W. (1945). Concentration regulation and volume control in *Lumbricus terrestris* L. *Nature, Lond.* **155**, 635.

STOUGH, H. B. (1926). Giant nerve fibres of the earthworm. *J. comp. Neurol.* **40**, 409–463.

STOUGH, H. B. (1930). Polarization of the giant nerve fibres of the earthworm. *J. comp. Neurol.* **50**, 217–229.

STRELIN, G. O. (1938). O fiziologicheskom gradiente III. Gradienty reduktsii vitali nykh krasor pri undushii u oligochaeta i ikh sviaz s povrezhdeneim. *Russk. Arkh. Anat.* **19,** 226–243.

SVEDBERG, T. (1933). Sedimentation constants, molecular weights, and isoelectric points of the respiratory proteins. *J. biol. Chem.* **103,** 311–325.

SVENDSEN, J. A. (1957). The behaviour of Lumbricids under moorland conditions. *J. Anim. Ecol.* **26,** 423–439.

SZYMANSKI, J. S. (1918). Die Verteilung von Ruhe- und Aktivitätsperioden bei einigen Tierarten. *Pflüg. Arch. ges. Physiol.* **172,** 430–448.

TANDAN, B. K. (1951). Axial gradient in the water content of the body wall of earthworms. *Current Science.* **20,** 214–215.

TAYLOR, G. W. (1940). The optical properties of the earthworm giant fibre sheath as related to fibre size. *J. cell. comp. Physiol.* **15,** 363–371.

THOAI, N-V. and ROBIN, Y. (1954). Métabolisme des dérivés guanidyles IV. Sur une nouvelle guanidine monosubstituée biologique: L'ester guanidoéthylsérylphosphoreque (lombricine) et le phosphagène correspondant. *Biochim. Biophys. Acta.* **14,** 76–79.

THOMAS, J. B. (1935). Über die Atmung beim Regenwurm. *Z. vergl. Physiol.* **22,** 284–292.

THORPE, W. H. (1956). *Learning and Instinct in Animals.* pp. 493. Methuen, London.

TÖRÖ, I. (1960). Research Report of the Department of Histology and Embryology, Medical University, Budapest. *Folia Biol., Praha,* **6,** 154–162.

TRACEY, M. V. (1951). Cellulase and chitinase of earthworms. *Nature, Lond.* **167,** 776.

TURNER, R. S. (1955). Relation between temperature and conduction in nerve fibres of different sizes. *Physiol. Zoöl.* **28,** 55–61.

TUZET, O. and ATTISSO, M. (1955). Migration des amoebocytes chez les Oligocètes terricoles. *C. R. Soc. Biol., Paris,* **149,** 798–799.

UMRATH, K. (1952). Über die Erregungssubstanz der sensiblen Nerven der Anneliden. *Z. vergl. Physiol.* **34,** 93–103.

VANNOTTI, A. (1954). *Porphyrins* (Trans. Rimington, C.). Hilger & Watts, London.

VOIGT, O. (1933). Die Function der Regenwurm Kalkdrüsen. *Zool. Jahrb. Abt. Allgem. Zool. Physiol. Tiere,* **52,** 677–708.

WALKER, J. G. (1959). Oxygen poisoning and recovery in the annelid *Tubifex tubifex. Dissertation absts.* **20,** 719.

WALTON, W. R. (1927). Earthworms and light. *Science,* **66,** 132.

WATANABE, Y. (1927). On the electrical polarity in the earthworm, *Perichaeta communissima* Goto et Hatai. *Tôhoku Imp. Univ. Sci. Rep.* 4th Ser. **3,** 139–149.

WATANABE, Y. and CHILD, C. M. (1933). The longitudinal gradient in *Stylochus ijimai;* with a critical discussion. *Physiol. Zoöl.* **6,** 542–591.

WATSON, M. R. (1958). The chemical composition of earthworm cuticle. *Biochem. J.* **68,** 416–420.

WATSON, M. R. and SILVESTER, N. R., (1959). Studies of invertebrate collagen preparations. *Biochem. J.* **71,** 578–584.
WATSON, M. R. and SMITH, R. H. (1956). The chemical composition of earthworm cuticle. *Biochem. J.* **64,** 10 P.
WELSH, J. H. and MOORHEAD, M. (1960). The quantitative distribution of 5-hydroxytryptamine in the invertebrates, especially in their nervous systems. *J. Neurochem.* **6,** 146–169.
WHERRY, R. J. and SANDERS, J. M. (1941). Modifications of a tropism in *Lumbricus terrestris*. *Trans. Illin. Acad. Sci.* **34,** 237–287.
WILLEM, V. and MINNE, A. (1900). Recherches sur l'excrétion chez quelques annélides. *Mem. Acad. Roy. Belg.* **58,** 1–73.
WILSON, D. M. (1961). The connections between the lateral giant fibers of earthworms. *Comp. Biochem. Physiol.* **4,** 274–284.
WITTICH, W. (1953). Untersuchungen über den Verlauf der Streuzersetzung auf einem Boden mit starker Regenwurmtätigkeit. *Schriftenreihe forstl. Fak. Univ. Göttingen,* **9,** 5–33.
WOLF, A. V. (1938). Studies on the behaviour of *L. terrestris* and evidence for a dehydration tropism. *Ecology,* **19,** 233–242.
WOLF, A. V. (1940). Paths of water exchange in the earthworm. *Physiol. Zoöl.* **13,** 294–308.
WOLF, A. V. (1941). Survival time of the earthworm as affected by raised temperatures. *J. cell. comp. Physiol.* **18,** 275–278.
WU, K. S. (1939a). On the physiology and pharmacology of the earthworm gut. *J. exp. Biol.* **16,** 184–197.
WU, K. S. (1939b). The action of drugs, especially acetylcholine, on the annelid body wall (*Lumbricus, Arenicola*). *J. exp. Biol.* **16,** 251–257.
YERKES, R. M. (1912a). Habit and its relations to the nervous system in the earthworm. *Proc. Soc. exp. Biol., N.Y.,* **10,** 16–18.
YERKES, R. M. (1912b). The intelligence of earthworms. *J. Anim. Behav.* **2,** 332–352.
ZHINKIN, L. (1936). The influence of the nervous system on the regeneration of *Rhynchelmis limosella*. *J. exp. Zool.* **73,** 43–65.
ZICSI, A. (1958). Freilandsuntersuchungen zur Kenntnis der Empfindlichkeit einiger Lumbricidenarten gegen Trockenperioden. *Acta. Zool. Acad. Scient. Hung.* **3,** 369–383.

# INDEX

Absorption 26
Acclimation 88
Accommodation of giant fibres 153
Acetylcholine 90, 97, 121, 136, 150, 154, 155, 156
Acid-base relations 26, 28–30, 35, 103
Acidity, responses to 179, 180
Acid phosphatase 56, 62, 110
Action potentials 161, 166
Active uptake of chloride 72
Adrenaline 90, 97, 121, 136, 156, 157
*Aeolosoma* 6
Aerobic metabolism 110
Aestivation 81–82, 131
After-discharge 148, 149
Albumen 8
Alkaline denaturation of haemoglobin 106
Alkaline phosphatase 31, 33, 57, 123
Allantoic acid 47
Allantoin 47, 49
*Allolobophora chlorotica* 3, 66, 71, 80
*Allolobophora longa* 2, 3, 11, 19, 20, 22, 50, 51, 57, 71, 73, 81, 82, 102, 103, 108, 121, 122, 129, 131, 149, 180
*Alma emini* 85
Amines 5
Amine oxidase 157
Amino acids 5, 61
Ammonia 46, 48, 49, 52, 53–68
Amoebocytes 98
Amylase 20, 22
Anaerobic metabolism 109, 123

Arginase 54, 55, 62
Arginine phosphate 1
Ash content 4
ATP 16, 116
Axial field 36–44


Behaviour 171–183
  development of CNS 172
*Bimastus eisenii* 22
Blood system 96–98
  blood pressure 65
  effect of drugs on 97
  effect of temperature 97
  flow to calciferous gland 32
Body wall, sensitivity to drugs 150
Bohr effect 85, 103
Branchiobdellidae 19
*Branchiodrilus* 84


Carbohydrates 4
Carbon dioxide 29, 30, 38, 95, 180
Carbon monoxide 100, 108
Carbonic anhydrase 29, 30
Calciferous glands 20, 24–35, 123
Calcium 2
Calcium carbonate 24, 25, 31, 32, 33, 34, 51
Cellulase 21, 22
Cerebral ganglion 129, 130–133
Chaetae 15, 144
Chemical composition 3
Chemoreception 166, 167, 173, 177
Chitinase 21, 22

# INDEX

Chloragogen cells   27, 55–68, 113, 121
  ammonia content   56
  urea content   56
Chloride, body fluid and urine   75–76
Cholesterol   4
Cholinesterase   90, 150, 156
Chromaffin cells,   156
Chromolipid   57
Cicatrice   58, 119
Ciliary actions in nephridia   66
Citrulline   55, 61
Cocoon production   132
Coelom
  hydrostatic pressure   81, 144
  water content   2
Colour forms of *A. chlorotica*   73
Conduction velocity
  giant fibres   158
Creatinine   54, 60, 62
Crop   19, 30
Cuticle   12–15, 70
Cyanide effect on respiration   36, 39
Cyclic activity   131, 132
Cytochrome-cytochrome oxidase   109, 123

Dehydration tropism   72
*Dendrobaena*   22, 73
*Dero*   36, 84
Desiccation   3, 77
Diameter and conduction of giant fibres   160, 161
Dissociation curves of haemoglobin   102
Diurnal rhythms   90, 91, 182
Dry weight   3

Effect of stretch on giant fibre   160, 161
Egg laying and neurosecretion   132
*Eisenia foetida*   38, 40, 42, 46, 50, 51, 81, 87, 88, 96, 108, 109, 120, 121, 122, 123, 129, 130, 131, 172, 180
  *rosea*   4
Elastin   15
Electrical gradients   38, 39, 119
Electrophysiology   150, 161
Eleocytes   66, 121
*Eophila*   94
Ephedrine   157
Ergosterol   5
Eserine   150, 156
*Eutyphoeus*   20

Facilitation of conduction   161
Fat content   4, 63
Feeding   177, 178
Filtration in nephridium   65
Food selection   18, 178
Freezing point depression of blood and coelomic fluid   73

Gaseous exchange   83–86
Giant fibres   135, 157–165
Gills   24, 84
*Glossoscolex*   89, 104
Gluconate-6-phosphate   110
Glucose-1-phosphate   110
Glucose-6-phosphate   110
Glutamic acid   55, 61
Glyceraldehyde dehydrogenase   122
Glycocol   61
Glycogen   42, 57, 109, 110, 121, 122
Glycolysis   42, 109, 122, 123
Gonadal maturity   131
Guanidoethyl-seryl-phosphodiester   16, 114
Guanine   59, 64, 65
Gut movements   154, 155
Gut pH   22, 23

Habituation   180
Haemochromogen   65, 106

Haemoglobin  85, 94, 95, 98–108
  alkaline denaturation  106
  dissociation curves  102
  effect of pH  103
  electrophoresis  108
  function of  98
  isoelectric point  104
  properties  104
Heteroxanthine  60
Homostrophic reflex  149
*Hoplochaetella*  63
Hormonal centres  43, 132
Hydration  70
Hydrostatic pressure  81, 144
Hydrostatic skeleton  71, 143
5-Hydroxytryptamine  121, 136, 156, 170
Hypotonic fluids on nephridia  66

Inanition  50, 52
Inhibitory voltage  40, 41
Intermediate metabolism  109–117
Intestinal flora and fauna  22
Intracellular recording  161, 162
Invertase  22
Iodine equivalence  38
Iodoacetic acid  112, 122
Ion uptake from exterior  72

Juvenile hormone activity  136

α-ketoglutaric acid decarboxylase  110
Krebs-Henseleit cycle  54, 61
Kynurenine  61

Lactic acid  42, 110, 112, 123
*Lampito mauritii*  2
Learning  180
Lichenase  19
Light
  effect on behaviour  174
  effect on CNS  166, 177
  regional sensitivity to  176
Limestone  27
*Limnodrilus*  36
Lipase  20
Lipids  23, 57
Lipoids  42
Locomotory activity  143, 144, 182
Lombricine  1, 5, 16, 114
  biosynthesis  115–116
Looped capillaries  84–85
*Lumbriculus*  36, 87, 156, 179
*Lumbricus*  2, 3, 4, 16, 18, 19, 20, 26, 30, 38, 46, 50, 51, 63, 66, 67, 71, 72, 73, 87, 88, 92, 102, 103, 107, 108, 110, 114, 123, 128, 131, 142, 146, 149, 155, 156, 161, 171, 174, 178

Malonate, effect on respiration  111
*Megascolex*  2, 63, 89, 159, 160
*Megascolida cameroni*  116
Metabolic gradient  41
Metabolic centre  41, 43
Migration  29, 73
Mineral salts  4
Mitochondria  31
Mucus  11, 45, 86
Muscle
  myosin  8
  protein  8
  structure  6
Muscovite  57

Naididae  173
Nephridia  64–68, 74, 77–82
  alkaline phosphatase  123
  changes in urine  78
  effect of external osmotic pressure  77, 80
  reabsorption  80
Nervous system  138–170
Neuromuscular junction  151–153
Neurosecretion  128–137

# INDEX

Nitrogenous excretion 45–68, 125, 126
Noradrenaline 157

*Octolasium* 22, 42
Omnivorous habit 18
Ornithine 61
Osmotic pressure 75
Oxygen debt 90

Paramyosin 9
*Peloscolex* 108, 109, 111
*Perichaeta* 4, 119, 142, 179
Peristalsis 142–145, 150, 168
Phagocytosis 58, 98
*Pheretima agrestis* 175
  *communissima* 8, 23
  *divergens* 23
  *hawayana* 89, 104
  *hilgendorffi* 38
  *posthuma* 2, 20, 65, 67, 73, 119
Phosphagen 5, 16
Phospholipids 57, 121
Photoreceptors 172
*Pontoscolex* 89
Post-synaptic potentials 163
Prostomium 144
Protease 20, 22, 26, 154
Proteins 4, 66
Protoporphyrin 9–11, 65
*Protoscolex* 104
Purines 57
Pyruvic acid 110, 123

$Q_{10}$ of respiration 88–89

Reflex arc 151–153
Regeneration 118–127
  effect of cyanide 124
  effect of drugs 120, 121
  electrical gradients 39, 119
  excretion 125, 126
  importance of CNS 118, 134
  number of segments 40, 122

Repetitive firing of giant fibres 162
Reproduction 132–133
Respiration 83–117
  cellular respiration 108–117
  effect of carbon monoxide 100, 108
  effect of cyanide 39, 108, 123
  effect of drugs 89, 90
  effect of partial pressure of oxygen 92–94, 124
  in regeneration 121
  rates of isolated tissues 36–38, 41
  rates of whole animals 87, 88
  rhythms 90
  tropical species 88, 89

Salivary glands 20
Secondary sex characters 131, 132
Sense organs 165–167
Sensory fields 141
Sensory input to giant fibres 159
Septa of giant fibres 163–165
Serine 116
Serine-diethyl-phosphate 114
Solid content 38
Sterols 4, 5
Stretch receptors 147, 149
*Stylochus* 39
Succinate 111
Succinoxidase 42, 122
Sub epidermal network 168
Sub oesophageal ganglion 129–135
Sulphur content 38
Surface area and respiration 93, 94
Survival time in waterlogged soils 71
Synaptic properties 151–153

Temperature
  effect on behaviour 172
  effect on CNS 169
  effect on blood system 97

Tension reflexes   146, 147
Thermal deathpoint   38, 42
Thigmotaxis   173
Touch receptors, active areas   142, 166
Transamidination   116
Transmitters   156
Transphosphorylation   116
Tricarboxylic acid cycle   111, 112
Tropomyosin   9
*Tubifex tubifex*   4, 19, 36, 70, 92, 93, 94, 95, 96, 109, 124, 171

Ultra-violet light   133, 176
Unloading tension   85
Urea   46–68
Urease   46
Uric acid   46–68

Urine
  hypotonic   70, 74, 77
  production rate   66, 81
  technique for collection   47

Ventral ganglia   129, 136
Vesicles at giant fibre septa   164, 165

Water
  content   27, 70
  loss   3, 71, 72
  regional differences   2
  relations   69–82
Weight of fresh animals   71

Xanthine oxidase   46
X-ray diffraction studies   13, 57